T0073546

Human Physiology: A Very Short Introduction

VERY SHORT INTRODUCTIONS are for anyone wanting a stimulating and accessible way into a new subject. They are written by experts, and have been translated into more than 45 different languages.

The series began in 1995, and now covers a wide variety of topics in every discipline. The VSI library currently contains over 650 volumes—a Very Short Introduction to everything from Psychology and Philosophy of Science to American History and Relativity—and continues to grow in every subject area.

Very Short Introductions available now:

Available soon:

For more information visit our website

www.oup.com/vsi/

Jamie A. Davies

HUMAN
PHYSIOLOGY

A Very Short Introduction

OXFORD
UNIVERSITY PRESS

OXFORD
UNIVERSITY PRESS

Great Clarendon Street, Oxford, OX2 6DP,
United Kingdom

Oxford University Press is a department of the University of Oxford.
It furthers the University's objective of excellence in research, scholarship,
and education by publishing worldwide. Oxford is a registered trade mark of
Oxford University Press in the UK and in certain other countries

Published in the United States of America by Oxford University Press
198 Madison Avenue, New York, NY 10016, United States of America

British Library Cataloguing in Publication Data
Data available

Library of Congress Control Number: 2020949491

ISBN 978–0–19–886988–7

Printed in Great Britain by
Ashford Colour Press Ltd, Gosport, Hampshire

Links to third party websites are provided by Oxford in good faith and
for information only. Oxford disclaims any responsibility for the materials
contained in any third party website referenced in this work.

Contents

Acknowledgements

First, I would like to thank Latha Menon of OUP for encouraging me to write this book in the first place. I am very grateful to members of my laboratory, and the Edinburgh medical community in which it is embedded, for very helpful discussions. I am, as always, grateful for Katie for her support, patience, and advice.

List of illustrations

Chapter 1
Human physiology: what it is and how it is done

The scope of human physiology

Human physiology is the science of how the body works. As a
science, physiology helps us to understand what our bodies are
and to enrich our knowledge of what it is to be alive, of what
it is to be human. As well as telling us how our bodies work,
physiological knowledge is important to keeping them working in
the event of injury or disease. Physiology is one of the four
intellectual foundations of modern medicine, the others being
anatomy, biochemistry, and pathology.

Anatomy and biochemistry focus on the material nature of the
body, anatomy concerning itself with parts and their arrangements,
and biochemistry with the molecules of life and how they interact.
The subject matters of anatomy and biochemistry exist whether
the body is living or freshly dead; that is why anatomists can
do their research by examining cadavers and biochemists can
perform useful analyses on liquefied extracts of tissues.

Physiologists, on the other hand, deal with processes that exist
only in the living, because they work not so much on what body
parts *are*, as on what they *do*. In that sense, physiology has a
greater claim to being the science of life than any other field.
Generally, physiology is taken to be the study of the workings of a

healthy body, the processes associated with ill-health and disease being the province of pathology. The boundaries between these two subjects are often indistinct, especially when they consider the interaction between diseases and the natural defences of the body (Chapter 7). There are also very strong links between physiology and biochemistry, because the functions and systems that underlie physiological actions depend ultimately on the interactions between the chemicals of which living things are made. Similarly, many physiological processes work only because tissues exist in a precise anatomical arrangement, and anatomy and physiology can often be understood only together. There will be several examples of this later in this book.

Physiology as a topic and as a method of working can be applied to any living organism and, while this book limits itself to human physiology, there are plenty of researchers who devote their lives to the physiology of trees or mushrooms or cuttlefish. In these organisms, too, they find much that is fascinating and important.

How human physiology is studied

As in all of the biological sciences, research in human physiology uses observation, inference, imaginative proposing of hypotheses, and the testing of these hypotheses by experiment.

Research often begins with observation, because it is usually only through becoming familiar with a phenomenon that a researcher can begin to frame a clear question. A simple example might be an observation that you may be able to make right now. Assuming you are reading this book while sitting quietly, use a watch or clock to count how many times you breathe in ten seconds, then count how many times your heart beats in the same time (see Figure 1 for a guide to taking your own pulse). Then subject yourself to moderate exertion, for example by running upstairs a few times, and repeat the counts.

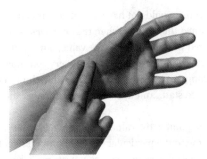

1. A simple way to take your own pulse. Gently place your finger tips across the thumb side of your wrist as shown in the diagram. With the fingers positioned as in the diagram, and your hand in line with your forearm (not flexed or flopping back) you should feel your pulse under your index finger.

You will find that your pulse rate and rate and depth of breathing will have altered. That is a simple physiological observation. Someone making such an observation might then wonder whether the two changes, in breathing and in pulse, are causally connected and might come up with a hypothesis about one being caused by the other. He might, for example, propose that faster breathing causes a faster pulse. This hypothesis is easy to test and again you can do it right now. Sit quietly again, for long enough that pulse and breathing have returned to normal after your temporary exertion. Then breathe rapidly for a short time (five seconds will be enough), and time your pulse again. Unless the experience of rapid breathing excited or frightened you in some way, which is unlikely, you will find that the fast breathing has not raised your pulse rate at all. Your experiment has therefore refuted the hypothesis.

The example in the foregoing paragraph, though very simple, is a real physiological experiment and illustrates several important points. One is the natural sequence of observation, leading to hypothesis, leading to experiment. Another is the value of making comparisons. Values for pulse rate or rate of breathing after running up stairs would not have meant very much had we not already obtained their values when the body was at rest. What really mattered was not the absolute value, but the change from sitting to stair-climbing or panting. It is very common, in all biology, to compare experimental data to something representing an unperturbed state, and this state is commonly called a 'control'.

A more subtle point is the value of planning an experiment that, if the result comes out a particular way, can prove a hypothesis wrong. This way of working may seem perverse, when we aim to establish what is correct, but it is powerful. Proving an idea to be wrong can usually be done with certainty, and this produces a useful gain in knowledge by eliminating ideas from the range of possibilities that have to be considered.

Proving an idea to be correct, on the other hand, is much more difficult. Imagine for a moment that the experiment had produced the opposite result, and breathing quickly while sitting had produced a faster pulse like the one seen when running up stairs. Would this have proved that breathing controls pulse? If your answer is 'yes', because the experimental result is compatible with the hypothesis, consider that the running up stairs and rapid breathing while sitting both involved wilful action. Someone championing a hypothesis that pulse is raised whenever wilful action takes place might have done exactly the same experiment, and the (imagined) experimental result would have supported her hypothesis just as much as the one about breathing controlling pulse. It is for this reason that experimental results that are compatible with a hypothesis do not generally prove that hypothesis to be right; there is usually a risk that they might support an alternative hypothesis as well, even if nobody has yet thought of it. Results that prove a hypothesis wrong give more certainty, and careful researchers therefore plan experiments that have the power to do this. The way that the experiment actually came out proved that raised breathing did not actually raise pulse, and eliminated at a stroke that idea from the set of all possible ideas about the control of the pulse rate.

A final point made by the exercise is that there is an important difference between correlation, the changing of one thing with another, and causation. The mere fact that two features A and B are observed together does not mean that A causes B: it might be that B causes A, or that A and B are independent consequences of some other cause (as is in fact the case for pulse and breathing: see Chapter 2). A failure to understand this point lies behind a vast quantity of misreporting of medical research in the popular press, especially of the kind that claims causal links between the life choices people make and the diseases they suffer. Some associations certainly do exist, that between smoking and lung cancer being the most famous, some seem definitely not to, such as vaccination against measles, mumps, and rubella and the development of autistic spectrum disorder. Most have simply never been subject to the rigorous testing that is always required to transform an observation of correlation to a clear indication of cause and effect.

Experimental models for physiology

Many experiments on human physiology are done directly on human volunteers: data obtained directly from humans are obviously of the highest relevance to human biology and, at least in the case of experiments that do no harm, they are free of serious ethical barriers. The simple experiment on breathing and pulse in the previous section of this chapter is an example, and so are many famous experiments in the history of physiology. William Harvey, one of the most important early physiologists, used a very simple experiment on his own body to refute the ancient idea that blood flows in both directions in arteries and veins. He tied a tourniquet around his arm, tight enough to block blood flow in the veins, but not so tightly that it could prevent blood flow in the

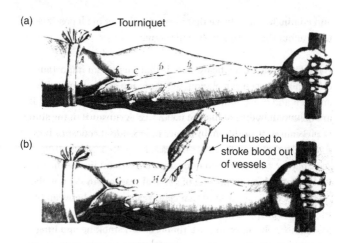

(a) → Tourniquet

(b)

Hand used to stroke blood out of vessels

2. William Harvey's experiment showing directional blood flow, shown using figures in his own book, *De Motu Cordis*: (a) shows the effect of an arm tourniquet, causing veins in the forearm to swell but veins in the arm above the tourniquet not to; (b) depicts a finger stroking blood back out of the veins towards the hand: the vein would not refill from the elbow end but would fill from the hand end when the stroking finger released its pressure. The implication was that blood flowed into the veins from the hand end back towards the body, and not the other way.

arteries, which have thicker walls and are harder to close with external pressure (Figure 2). He watched as the veins below his tourniquet swelled up with blood, while the veins above the tourniquet remained empty. Furthermore, he could stroke blood out of short lengths of his veins using his fingers: it re-entered only if allowed to from the hand end, not from the body end.

The implication was obvious (once someone had the idea of doing the experiment in the first place): blood flows in one direction only, entering the arm through the arteries and leaving it through the veins. Harvey reasoned that there must be fine blood vessels, too small to see, that connected the end of the arterial system deep in the tissues to the beginnings of the vein system. These blood

6

vessels, capillaries, had not been observed directly and would not be for another forty years but, by assuming they were there, Harvey laid the foundations of the theory that blood circulates in one direction from the heart, in arteries to the tissues, through capillaries, and then back to the heart in veins. He did not have all of the details correct in his book, but his simple experiments on a living human body had made a considerable advance in the study of physiology. In addition, by showing that ancient ideas of how the body worked, handed on from classical times, could be wrong, Harvey's simple experiment spurred other researchers into putting their antique books away and performing experiments of their own.

Many questions of physiology cannot be answered using healthy humans because they concern internal processes that cannot be accessed from outside. Exceptionally, however, specific injuries and diseases have allowed doctors and surgeons to learn about normal human physiology in the course of treating their patients. A classic example is provided by the case of Alexis St Martin, a Michigan fur trader who was shot in the stomach in 1822. His wound healed in an unusual way, the edges of the stomach wound uniting with those of the skin wound to leave a fistula (a hole) through which his doctor, the army surgeon William Beaumont, could introduce food to the stomach on the end of a string and withdraw it again at different times, to study the digestive process. He could also withdraw stomach liquids to study their effect on food outside the body. His work showed for the first time that digestion was a chemical process, not a mechanical one. St Martin, who presumably gave his consent to these experiments, went on to live for several more decades.

Surgeons have always been able to observe internal physiology during operations, although generally they have not been able to conduct experiments to correlate stimuli and response (at least, not within the realms of normal ethical science). There are, though, some exceptions: brain surgeons sometimes operate on conscious individuals, who are sedated as the surgeon gains access to the brain but allowed to regain consciousness when she has reached there and is preparing to operate. The reason that the patients are conscious is that the surgeon can gently stimulate parts of the brain and ask patients to perform simple tasks like raising a finger, or to report what they feel. By listening and observing the patients, the surgeon can confirm exactly which part of the brain she is touching, and this helps the precise planning of an operation so that it does no serious damage. While this is done for purely medical reasons, the more it is done, the more is learned about the precise relationship between mental functions and specific locations on the surface of the brain.

Imaging technologies, most of which were developed in the 20th century, have greatly expanded the range of research that can be done on people. X-rays, discovered by Roentgen in 1895, were quickly applied to anatomical and physiological studies showing, for example, how bones move in relation to one another when a limb or digit is moved. Indeed, I am old enough to remember a rural shoe shop that had an ancient pedoscope, an X-ray machine in a large wooden cabinet with a hole for a foot, which was designed to see the quality of fit of a shoe and which was a common device in the 1930s and 1940s. I can well remember wiggling my toes in such a machine and watching my infant bones moving. In 1970, it must have been one of the last of these thoroughly dangerous machines still in use. To my disappointment as a child, but retrospective relief as an adult, it had disappeared before my next visit to that shop. Early in the X-ray era, people's bodies were observed under X-rays as they moved, breathed, swallowed, and performed other ordinary actions, allowing researchers to have unprecedented views of the internal motions of the body (this is still done for medical investigations, but with very much lower and safer doses).

Ultrasound imaging allows much safer observation of both hard and soft tissues and, in a form called Doppler ultrasound, it can reveal features such as blood flow. Blood flow can also be tracked, at fairly high resolution, by a technique called functional magnetic resonance imaging (fMRI), a technique that uses blood flow to detect which areas of a living brain are working hardest. This type of imaging, developed in the 1990s, can be done on healthy humans without discomfort. It has made it possible for researchers to observe people as they are presented with stimuli or tasks and has allowed them to associate different types of mental activity with different brain regions.

Some physiological information can be inferred from external measurements. The body's use of oxygen, its pulse rate and pressure, its temperature and its muscle power can all be measured fairly easily in appropriately modified gymnasia, and such measurements are valuable to physiology and one of its applied daughter fields, sports medicine. Portable measurement equipment can also be used to measure human responses to unusual environments, for example the high acceleration forces of the launch stages of space flight and the 'zero-g' free-fall conditions of orbit.

There are, though, limits to what can be studied directly in humans. For this reason physiology has, for all of its long history right back to ancient Greece, featured experiments on living non-human animals. No introduction to physiological methods would be complete without a clear acknowledgement of this fact. Animal experimentation has been, and continues to be, highly controversial, mainly for ethical reasons, though there are also some technical arguments. The ethical arguments revolve mostly about whether causing deliberate harm to one being (e.g. a dog) can be justified either by the desire to gain knowledge as an end in itself, or by the hope that medical application of that knowledge will do good to other beings (humans or, through veterinary medicine, other dogs). Details of the ethical arguments are beyond the scope of this book but it is important to note that scientists themselves are not only aware of the issues, but generally feel the same emotional responses to animal suffering and death as anyone else would. Studies on those who are involved in the use of laboratory animals have reported elevated incidences of psychological difficulties caused by the work, including depression, substance abuse, and post-traumatic stress disorder. The great physiologist Claude Bernard (see Chapter 3), who experimented on living animals extensively, wrote in 1865 that 'the science of life is a superb and dazzlingly lighted hall which may be reached only by passing through a long and ghastly kitchen'. Eighteen years later, Bernard's wife, who had presumably seen parts of that 'ghastly kitchen' for herself, founded France's first society dedicated to campaigning against animal experimentation.

Darwin wrote of animal experimentation, 'I quite agree that it is justifiable for real investigations on physiology; but not for mere damnable and detestable curiosity. It is a subject which makes me sick with horror...' He also promoted a movement to place animal experimentation under strict legal control so that Parliament and its appointed inspectors could decide what was, and what was not, acceptable in the name of science. The result was the UK's Cruelty to Animals Act (1876), the world's first legislation designed to regulate scientific use of animals. The Act was an ancestor to the Animals (Scientific Procedures) Act (1986) that regulates current animal use in the UK; in most other developed countries too, animal experimentation is allowed, but only by trained and licensed scientists working in licensed premises and performing work that has been approved by an independent body. This legislative framework is a compromise that satisfies some people, but not all, and most scientifically active countries have one or more pressure groups working to outlaw any use of living animals in scientific research.

The technical objections to the use of animals to study human physiology rest mostly on the question of how realistic are the animal models ('model' in the sense of 'representation'). The answer generally depends on context. Much basic physiology is

10

shared across all mammalian species, meaning that what is learned in one species can be carried across to another, but there are still differences. Anyone who keeps a dog will have observed that the basic processes of breathing, eating, excreting, and sleeping seem to be very like those of humans. Also, when we are cold, humans and dogs both shiver. When we are hot, however, differences in our physiology become apparent. We humans lose heat by behavioural changes (reducing activity, seeking shade) and by sweating, which sheds heat in vapour evaporated from damp skin. Dogs use the same behavioural changes but they do not sweat (except a tiny amount in their paws), presumably because their body hair would make it ineffective. Instead, they pant, using rapid, very shallow breaths to pass air over their large wet tongues to shed heat into departing vapour. The cooling responses of the two species both make use of the same principle of evaporating a fluid, but the precise solution is adapted to our different types of body surface.

The difference between sweating and panting is obvious, but internal differences can be more subtle. They have often led to drugs tested to be safe in animals being found to be dangerous to humans, or dangerous to other animal species. Another limitation of animal experiments is that some of the features of humans in which we may be most interested are, as far as we can tell, absent from other animals. The brain function that allows you to read and understand this sentence is an example.

Allowing for the fact that there are exceptions, we are still left with the general principle that most aspects of human physiology work in an identical, or very similar, manner in other mammals. Therefore, ethics and law permitting, it is possible for researchers to learn a great deal about the working of human bodies by experimenting on animals in ways they would never be allowed (or wish) to do on humans. A typical experiment will test a hypothesis about a mechanism by deliberately interfering with the mechanism in some way. Many of these experiments do cause harm, often so much harm that they are done only on anaesthetized animals that are never brought round and are given a terminal dose of anaesthetic at the end of the work. Most of what we now know, including most of what is in the following chapters, comes from research that has involved animals at some stage, even if it was later verified in humans. With rather few exceptions, it is fair to say that the knowledge could not, at the time, have been obtained any other way.

Will animal experiments always be needed? In 1959, the zoologist William Russell and the microbiologist Rex Burch published *The Principles of Humane Experimental Technique*, in which they set out three goals for the research community. Now summarized as 'the 3Rs', the goals were to Refine animal experiments (make them less harmful), Reduce the numbers of animals used, and Replace animal experiments with alternatives. The 3Rs have become an important strand of science. In the UK, for example, researchers are legally required to consider them when planning experiments and the government funds a National Centre for 3Rs, which sponsors research into methods to refine, reduce, or replace animal experiments. Commercial bodies also take these issues seriously, and indeed some companies exist specifically to produce equipment and services for animal-free methods.

Work towards refining, reducing, and replacing animal use is currently dominated by three approaches. One is the use of 'lower' animals, such as fruit flies, in place of 'higher' ones such as mice: this approach works well for the physiology of cells and tissues but is clearly not helpful for physiology of, for example, temperature control (flies are 'cold-blooded'). Another is the use of human cells and human organoids—tiny and immature versions of human organs—that are grown in dishes in the lab. Again, these are excellent for study of physiology at very local levels but they are not useful for studying interactions between the organ in question and the rest of the body. A third approach is to use our existing knowledge of physiology to create abstract models of the living body in computers. Computer modelling can be very powerful, especially in helping to turn a vague set of questions and ideas into precisely formulated hypotheses, but it is limited because we do not yet understand enough about the body to be able to capture it in computer code. Many misleading statements have been made about the power of these three techniques to replace all animal experiments: none is anywhere near developed enough to do so.

At present, then, and for the foreseeable future, human physiologists continue to work directly on humans when possible, and they use human-derived cells and organoids to study internal processes whenever they can be of use, but they still rely on legally permitted animal experiments for what cannot be answered any other way. Each individual, whether scientist or not, will have their own ethical stance on this. This author has never performed experiments on living higher animals but he has certainly used results others have obtained this way to plan his research, and has also used these results in the writing of this book.

The nature of explanations

Different areas of science each have their own ways of explaining the world, and their own styles of recording and presenting an important discovery. A theoretical physicist, for example, might encapsulate knowledge in the form of a mathematical equation (e.g. $E = mc^2$). A chemist might use a chemical equation (e.g. $2H_2 + O_2 \rightarrow 2H_2O$) or a diagram that represents the spatial arrangement of atoms in a complex molecule. An evolutionary biologist might draw a phylogenetic tree whose branches show how different types of animal diverged over aeons of geological time. Each style of explanation has grown up to suit the aims and culture of its community and the styles most favoured by one can be quite useless to the other. How does a typical explanation in physiology look?

The very breadth of human physiology, which can be pursued from the level of molecular interactions to that of a whole person, means that several styles of explanation may be encountered. Nevertheless, it is probably fair to say that two dominate, one in scientific communication and the other in descriptions aimed at non-specialists, for example in television programmes or museum labels. The style of explanation most common in scientific discourse on human physiology is a graph, using that word in its most general sense—a set of points and lines. Such a graph, an example of which is shown in Figure 3, usually depicts physiological entities (e.g. cells, tissues, organs), appearing in words or as small diagrams, and it connects them with arrows that represent interactions (e.g. signals carried by hormones). The connectors obey conventions, an arrow-head usually meaning a positive influence (this hormone increases an activity of the receiving tissue), and a 'T' end meaning an inhibitory influence. A 'T' pointing at an arrow rather than at a tissue implies that the signal carried on the 'T' inhibits the signal carried by the arrow.

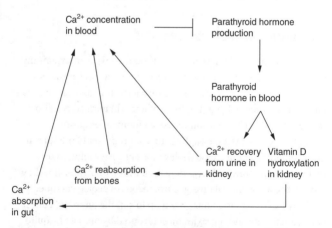

3. **An example of a graph depicting a physiological process, in this case features of the feedback loop by which adequate calcium (Ca^{2+}) ions in the blood inhibit parathyroid hormone production, which would otherwise drive a variety of mechanisms to increase blood calcium levels.**

There are no strict rules about these diagrams, and it is common for them to deal with entities at completely different scales, a cellular conversation in one part of the diagram influencing a whole organ at another. The diagrams are also not in themselves quantitative, being designed to give an intuitive idea of how something works rather than to be a predictive tool. For quantitative prediction, the diagram may be associated with equations every bit as mathematical as those of a physicist.

The point about these diagrams, which will be found in subsequent chapters of this book, is that they present the idea of a *system*; a set of components interacting in specific ways to generate properties at the system level that are absent from any one component. The existence of system-level properties that are not seen in their components is called emergence, and is a very important feature of living systems. Indeed, life itself can be seen as an emergent property of the physiological mechanisms that underlie it.

Physiological public discourse often also relies on analogy between a mechanism of the body and a mechanism familiar from the everyday world. It is noticeable that, over the history of physiology, these analogies have tended to draw parallels with the cutting-edge technology of the day. In the early industrial revolution, the heart and lungs were seen as types of pump; in the 1930s the nervous system was like a telephone exchange; in the 1980s the brain was a kind of computer; right now, countless physiological entities are described as networks or webs of information. Analogies such as these can be useful but they must not be taken too far: to insist that a heart is just a pump or the brain just an internet made of neurons will be to miss important features of the living that are absent from our machines. It is for this reason that this book will generally avoid using the traditional explanation-by-analogy, and will instead focus on living mechanisms themselves, however strange they may seem.

Chapter 2
Energy: food, oxygen, blood, heat, waste

Energy is essential for life

Life of any type is impossible without a supply of energy. As in the world of machines, all movement other than falling, or drifting with wind and water currents, requires energy, and so does remaining warm in a cool environment or cool in a hotter one. Energy is also needed for a more subtle reason. The constituent components of a living being are in a much more organized and ordered state than they are outside the body. Consider the intricate architecture of, say, a leaf viewed at any magnification, and contrast it with the materials from which the molecules of the leaf came; unstructured gases of the air and watery salt solutions in the soil. Creating ordered (low-entropy) structures from disordered (high-entropy) precursors, and maintaining the ordered state against forces that would dissipate it, cannot be done for free: it requires energy. Just as with man-made structures, if energy is not expended to maintain them, the order of living systems is gradually lost and they decay until they are no longer recognizable.

The ultimate sources of energy for life on earth are nuclear reactions. For some specialist organisms deep in the earth's crust or near undersea volcanoes, the most relevant nuclear reactions are those that continue to keep the core of the planet hot, the mantle flowing, and heat and minerals moving in the crust. For most of life, the most relevant nuclear reactions are those that take place in the sun and bathe the surface of the earth with warmth and light. Plants and photosynthetic bacteria make direct use of this energy, using light to power chemical reactions that combine simple molecules such as carbon dioxide and water to synthesize the more complex molecules from which living tissues are made. All other organisms acquire energy indirectly, either by consuming plant material that has been built by direct use of sunlight, or by consuming animals that have themselves eaten plants or have eaten other plant-eating animals. Most humans eat both plant and animal-based foods: living on non-animal food is possible, but only if great care is taken to find sources of vitamins normally found only in animal products. Food can be used both to provide raw materials for building new tissues, and for energy. This chapter will consider energy only; the topic of tissue-building is covered in Chapters 3 and 8.

Most of the biochemical reactions that underlie life happen insignificantly slowly unless they are facilitated by an enzyme (a biological catalyst). Living cells contain thousands of different enzymes, each of which catalyses a specific reaction. Many reactions group into 'pathways', in which a starting material A is turned into product B by enzyme 1, and product B into products C and D by enzyme 2, and so on. Typically, the 'entrance enzyme' of each pathway is controllable by the influence of a specific molecule, sometimes another enzyme, and this allows a cell to control how active a pathway should be at any one time. This control will be discussed further in Chapter 3.

All chemical structures contain a certain amount of energy locked in to the bonds between their constituent atoms. Chemical reactions run 'downhill' in energy terms, such that the total energy locked up in reaction products is less than that locked up in their starting materials (including light as a 'starting material' if, as in plants, light is relevant to the reaction). Petrol and oxygen, for example, together have much energy locked up in their structures; when they react together, they make relatively low-energy carbon dioxide and water, allowing a motorist to harvest the energy difference and apply it to moving a car (Figure 4(a)). Many biochemical reactions are like this, in that they begin with a high-energy molecule and produce a lower-energy product. Demolition of large molecules into their smaller components, as when food is used for energy, typically works in this direction (at least, overall). But cells also have to construct complex molecules from simple precursors, and perform various other reactions in which the 'desired' final product has more energy locked up in it than its starting component. This 'uphill' direction of a chemical reaction is possible only if it is coupled to a larger 'downhill' reaction, so that the total energy flow is still downhill (Figure 4(b)). The cell therefore has to pay for locking up energy in a new product, by using energy it has already locked up in an energy-donor molecule.

By far the most common energy-donor molecule is adenosine triphosphate (ATP), which can be split in a downhill reaction into adenosine diphosphate (ADP) and free phosphate, liberating energy that can be passed to an uphill reaction. Like a storage battery, the ADP can be 'recharged' by adding the phosphate back, as long as this uphill reaction is coupled to an even larger downhill reaction of another kind. The amount of energy available when ATP becomes ADP and phosphate is about 0.3 electron-volts (eV); this is equivalent to about 12 million-billion-billionths of a kilocalorie, the unit in which energy content of food is usually measured. ATP is such a useful energy-donor molecule that life has evolved to have uphill chemical steps that are remarkably standard in their energy requirements, most needing the 0.3eV of a single ATP. Some, such as that in Figure 4(b), need a small multiple of that. The problem of supplying the living cells of the human body with energy is therefore essentially one of obtaining energy from outside the body and using it in a way that liberates in 0.3eV packets to 'recharge' spent ATP.

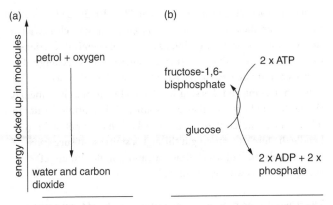

(a)

energy locked up in molecules

petrol + oxygen

water and carbon dioxide

(b)

fructose-1,6-bisphosphate

2 x ATP

glucose

2 x ADP + 2 x phosphate

4. **Energy flows in chemical reactions.** (a) Many reactions, such as the combustion of petrol shown here, run 'downhill' in energy terms, the final molecules having less energy locked up in them than the starting materials. (b) Where cells need to produce a final molecule with more energy locked up in it than its constituents, they couple the uphill reaction to a downhill one, so that, taking the coupled reaction into account as well, the total direction is still downhill.

Burning without flame

The simple experiment of placing an item of energy-rich, dry food, such as a glucose tablet, on hot coals and watching it flare up into hot flames, or watching a sugar-powered rocket take off, provides a vivid illustration of how much energy can be liberated in the form of heat when food is burned. Burning works by oxidation; glucose molecules, which have a moderately complicated arrangement of carbon, oxygen, and hydrogen atoms, combine with oxygen from the air and rearrange their chemical bonds to produce simple three-atom products (carbon dioxide and water), and liberate the energy locked up in the glucose as heat. The body uses essentially the same reaction—add oxygen and turn sugar into carbon dioxide and water—but it needs to harvest the energy to its ATP storage systems rather than to liberate it all as heat. This harvesting is made complicated by the fact that oxidation of glucose liberates about 29eV per molecule, a hundred times more than can be accepted by a recharging ATP molecule. Cells therefore run the oxidation not as a single event that releases all of the available energy at once, but as a series of chemical reactions that release energy bit by bit, transferring the released energy of each step ultimately to charge ADP to ATP. The difference between burning and this step-by-step metabolism is, in principle, similar to that between allowing a source of high-altitude water to fall as a single waterfall on to a waterwheel that cannot possibly handle the force, or running it through a staircase of successive waterwheels, each of which draws an appropriate amount of energy from its own step (Figure 5).

The reactions themselves are more the province of biochemistry than of physiology but, in outline, the simple 6-carbon sugar, glucose, is chemically modified, rearranged, and cleaved into two 3-carbon molecules. The first step of modification happens to be the one shown in Figure 4(b), and is an uphill reaction requiring the investment of energy from ATP before the later downhill reactions begin. This feature is typical of biology, in that the metabolism to run life requires life already to be running and, whatever happened billions of years ago when life first began, modern organisms cannot 'bootstrap' themselves from the non-living. Instead, they must develop from eggs or spores richly supplied with molecules such as ATP by the previous generation (see Chapter 8).

The 3-carbon sugars produced by modification and cleavage of glucose go through a series of chemical reactions (glycolysis), some of which release small amounts of energy that are transferred to energy-storing molecules called NADH and ATP. The end product of glycolysis, a 3-carbon molecule called pyruvate, can be used for several different chemical pathways but the most famous one, in terms of energy production, is the tricarboxylic acid cycle (TCA cycle, also known as Krebs' cycle). This transfers the energy of each pyruvate to four molecules of NADH and one of the similar molecules, $FADH_2$, while liberating

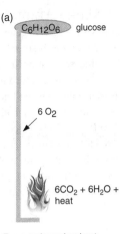

(a)

$C_6H_{12}O_6$ glucose

$6 O_2$

$6CO_2 + 6H_2O +$ heat

Energy released as heat

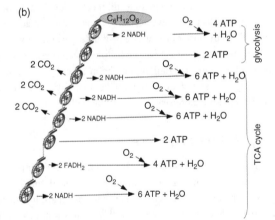

(b)

$C_6H_{12}O_6$

$\rightarrow$ 2 NADH O_2 4 ATP $+ H_2O$ glycolysis

$\rightarrow$ 2 ATP

2 CO_2

$\rightarrow$2 NADH $\rightarrow$ O_2 6 ATP $+ H_2O$

2 CO_2

$\rightarrow$2 NADH $\rightarrow$ O_2 6 ATP $+ H_2O$

2 CO_2

$\rightarrow$2 NADH $\rightarrow$ O_2 6 ATP $+ H_2O$

$\rightarrow$ 2 ATP   TCA cycle

$\rightarrow$2 FADH$_2$ O_2 4 ATP $+ H_2O$

$\rightarrow$2 NADH $\rightarrow$ O_2 6 ATP $+ H_2O$

Energy from each glucose molecule captured and stored in 36 ATP molecules

5. Two ways of 'burning' glucose: (a) shows simple chemical burning, as when a glucose tablet is put on a fire and the energy is released in one step, as heat; (b) shows typical biological use, where the energy of the glucose is harvested in stages, most of it being transferred to the cell's energy storage molecule, ATP. The energy storage molecules mentioned are not products of the glucose being used, but cycle round in the cell, being 'charged' with energy in the process above and 'discharged' when they give their energy up where it is needed.

three carbons as carbon dioxide. The NADH and FADH$_2$ later undergo somewhat indirect reactions with oxygen to generate water, and energy liberated from those reactions is stored in yet more molecules of ATP. Just as when glucose is simply burned, one molecule of glucose and six molecules of oxygen turn into six molecules of carbon dioxide and six molecules of water but, when the glycolysis and TCA cycles are used, the energy is not all liberated as heat alone but a significant part of it is stored in 36 molecules of ATP. Together, these 36 ATPs contain about 11.4eV of captured energy, an overall efficiency of harvesting the 29eV locked up in glucose of around 40 per cent; the other 60 per cent appears as heat. This efficiency is similar to that achieved by a modern diesel engine.

The energy-harvesting chemical reactions described above happen (sometimes with variations) in all human cells. For them to be possible, these cells must have access to glucose and oxygen, and must be able to dispose of the waste products, carbon dioxide and water. In a large organism such as a human, even these apparently simple requirements are challenging and are met by a complex set of physiological systems, in particular the digestive, respiratory, cardiovascular, and excretory organs of the body.

From grapefruit to glucose

The chemical reactions described above began with a simple, small molecule, yet most food is chemically and structurally very complex, being either still-living tissue, such as an apple plucked from a tree, or once-living tissue that may or may not have been cooked. Reducing food to simple molecules and delivering these molecules to the body is the main task of the digestive system, essentially a long tube that runs from mouth to anus, and which has a few side branches to specialized accessory organs such as the pancreas and liver (Figure 6). Consider a classic breakfast dish, half a grapefruit. This still-living plant tissue consists of water, simple carbohydrates such as sucrose, complex carbohydrates

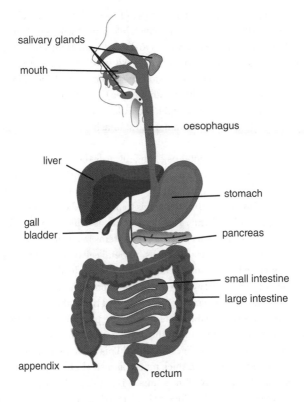

6. The human digestive system. For the sake of clarity, the small intestine has been shown much shorter and less convoluted than it really is, the liver and gall bladder have been depicted pulled free of the intestines to which they connect, and the height of the chest has been foreshortened.

such as cellulose, and small amounts of protein and fats. Its passage of the digestive system begins at the mouth, an organ that has teeth to cut and crush food to make its components much more accessible to later chemical attack. The dental enamel that forms the surface of teeth is the hardest of animal substances and can therefore cut through most types of food. The mouth is also equipped with salivary glands, which lubricate it with a watery fluid that contains salts, proteins to defend against infection, and some digestive enzymes such as amylase, which breaks starch down into simple sugars. Once food has been chewed enough to be semi-liquid, it is swallowed and passes through the pharynx to the oesophagus, a long, straight, muscular tube that runs down the middle of the chest from neck to stomach.

The stomach is a muscular bag lying towards the left side of the abdomen. Empty, it is about 20 cm long but its inside is highly folded to allow it to expand greatly to accommodate a large meal. The main function of the stomach is to break food down into a thick liquid called chyme. This liquefaction is achieved mainly by the digestion of proteins in the food, proteins being critical to the solidity of tissues. Specialized cells (parietal cells) in the stomach lining secrete hydrochloric acid, making the inside of the stomach very acidic, while other cells (chief cells) secrete the acid-tolerant, protein-digesting enzyme, pepsin. Having a tissue that secretes acid and enzymes capable of destroying the proteins that are the basis of tissue structure raises the obvious problem that the stomach is in danger of digesting itself. This is avoided by the stomach lining itself with thick mucus as a shield from its own secretions. The activity of the stomach is regulated both by anticipation (the smell of food being cooked) and by the stretching of the stomach wall by food. Although it does so much to break food down, the stomach is not strictly necessary for digestion and people who have had to have their stomach removed have only minimal digestive problems. They do require injections of vitamin B12, though, because the stomach is the only source of a factor needed for this vitamin to be absorbed later.

Beyond the stomach is the small intestine. This is 'small' in diameter but not in length, being about 4 metres long in life and twice that in death, once natural muscle tone is lost. The small intestine has various specialized zones but, overall, it is

24

responsible for digestion of food into simple constituent molecules, and for passing these molecules into the body. It accepts food from the stomach and, in addition, it is supplied with digestive juices from the pancreas, and bile salts. These bile salts contain biological detergents, useful for emulsifying fats. The liver also secretes some waste products, notably bilirubin from the breakdown of old blood cells, into the intestines for disposal. The enzymes and bile entering the gut from these accessory organs are responsible for most food digestion, which renders most complex molecules, such as proteins, starches, and fats, down into their small amino-acid, sugar, and triglyceride building blocks. Digestion is generally by hydrolysis: the breaking of a chain of a polymer and the capping of the broken bonds of the polymer with components of water (a hydrogen atom on one broken end, an O-H unit on the other). An example reaction is the hydrolysis of the two-sugar molecule, sucrose (the sugar most widely used in domestic cooking), into its subcomponents, glucose and fructose (Figure 7). The glucose is, of course, the same molecule that can directly enter the energy-harvesting biochemical systems described above, at least once it has been brought to the cell that needs it. Digestion is aided by churning mechanical contractions in the gut.

The inner lining of the intestine has a huge surface area for the absorption of food (Figure 8): this area is achieved first by the lining being folded into an alternating series of ridges and valleys a few millimetres across, then by the surface of these valleys having many tight-packed columns, the villi, projecting from it, about 1/10 millimetre across and 1 millimetre high. Each cell on the surface of a villus itself projects tiny columns, microvilli, each around 1/1000 millimetre across and bearing digestive enzymes, into the food-bearing fluid (the exact shapes and dimensions vary in different parts of the small intestine). These cells at the surface of the villi are specialized for absorption of nutrients, and their membranes have specific transport proteins that take molecules such as sugars and amino acids (the subunits from which proteins are made) into the cell from the gut fluid and expel them from

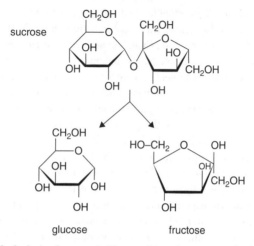

7. **The hydrolysis of sucrose (table sugar) into the monosaccharides glucose and fructose. This type of reaction, in which subunits are separated, with one end of the broken bond acquiring an H and the other an -OH, both from water (H_2O), is very common in the digestion of a large variety of molecules.**

their lower surfaces. Here, they enter vessels carrying blood, which is where most of the glucose will go, or lymph, to which components of fats, especially, go. As well as absorbing subunits of food, the small intestine also reabsorbs bile salts, which are recycled for later use. Other cells, between the villi, make mucus and defensive proteins, and make new absorptive cells to replace ones that have become worked out and lost.

After spending about two hours in the small intestine, the unabsorbable components of food, together with debris from damaged villi and remaining digestive juices, are expelled into the large intestine, only about 1.5 metre long but much wider than the small intestine. The large intestine is dominated not by human cells but by bacteria, which outnumber human cells by around a hundred to one. These bacteria, a complex mixture of species, have enzymes that human cells do not, and they can digest

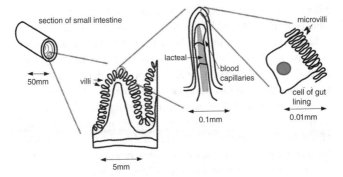

section of small intestine

microvilli

lacteal

blood
capillaries

villi

cell of gut
lining

50mm

5mm

0.1mm

0.01mm

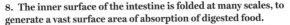

8. The inner surface of the intestine is folded at many scales, to generate a vast surface area of absorption of digested food.

complex carbohydrates such as cellulose that could otherwise not be digested at all. They also make vitamins (e.g. many B vitamins and vitamin K). Bacterial metabolism is also responsible for the generation of intestinal gases (flatus), including odourless hydrogen, methane, nitrogen, and carbon dioxide, and far-from-odourless compounds such as dimethyl sulphide. Bacteria in the large intestine also turn bilirubin from the liver into urobilinogen, the brown pigment responsible for the colour of faeces, which would be pale grey without it. The large intestine reabsorbs much water from the remains of the food which, as it dries and compacts under muscular action of the large intestine, becomes faeces, and moves the faeces to the final part of the gut, the rectum. Food normally spends between half a day and a day in the large intestine; fast passage caused by irritation of the intestine wall results in too little water being reabsorbed from the faeces, and therefore diarrhoea. Very slow passage, caused for example by a fibre-poor diet, causes absorption of too much water, over-solid stools, and constipation.

The blood vessels that leave the intestine lead directly to the largest internal organ of the body, the liver. Enzymes in this organ turn simple sugars other than glucose, such as fructose and galactose, into glucose (the liver does a lot else besides this). Glucose that is not used by the liver itself passes out of the organ in blood to circulate round the body as a whole. Liver and muscle cells take up glucose well and, if they take up more than they need for their own requirements for ATP generation, they combine glucose molecules into a storage polymer, glycogen. Glycogen accounts for around 15 per cent of energy storage in a healthy body. Most of the rest of energy storage is in the form of fat, some of which comes directly from the diet and some of which is made from excess glucose; that is why someone can become fat even on an almost fat-free diet. When glucose levels are low, fat can be used for ATP production; indeed, some cells prefer to use fat for their energy anyway. The efficient use of fat for energy depends on there being some glucose present; in its absence, the body has to send products of fat metabolism through a biochemical pathway that results in the production of 'ketone bodies', responsible for the fruity smell characteristic of the breath of people who are starving, either through lack of food, or because they have uncontrolled diabetes.

It must be stressed that this short description of the digestive system has focused, for simplicity and brevity, on how the body obtains glucose for energy. Humans need a range of different nutrients to maintain themselves, and a diet very high in simple sugars, which is becoming all too common in many countries, is a recipe for developing malnutrition and obesity at the same time.

Oxygen

For a land animal such as a human, the only source of molecular oxygen is the air, a fifth of which consists of that gas. Using this abundant resource is, however, not trivial. Oxygen is only poorly soluble in water and the oxygen demands of mammalian tissues are high. For a bulky animal like a human, it is therefore essential that oxygen be transported deep inside the tissues so that no cell is very far from a rich source. Given the vast number of cells in a human (about ten trillion), it is also essential that the surface through which oxygen can be absorbed is immense—much bigger than the skin. These two problems are solved by passing air across a large area of wet surface, packed into the lungs, and by using a blood system equipped with specialist oxygen carriers to carry absorbed oxygen into the tissues.

The lungs consist of branching air tubes (bronchi and bronchioles), with each branch ending in a tiny balloon-like sphere, the alveolus. There are around a quarter of a billion alveoli in each lung and their total area is around 75 square metres. The air-facing surface of the alveolus is wet, and this moisture is kept spread across the surface of the alveolus, rather than beading up, by a natural detergent-like molecule, lung surfactant. The wall of each alveolus is very thin, and fine blood vessels—capillaries—are situated immediately below. The path from air to blood is therefore very short, and oxygen can flow efficiently to the blood.

The problem of oxygen dissolving only poorly in water (the main constituent of blood) is solved by the presence of haemoglobin, the molecule that gives the cells that contain it the name of 'red blood cells' (erythrocytes) and gives blood its characteristic colour. Haemoglobin is a complex of four globin protein chains and an iron-containing organic molecule called haem. The manner in which haemoglobin binds oxygen is strange and subtle, and understanding it took a great deal of research in the mid-20th century. Once one molecule of oxygen binds to haemoglobin, the molecule changes shape, allowing it to bind another two molecules of oxygen more easily, and when these are present a fourth molecule of oxygen can bind more easily still. This cooperative binding of oxygen to haemoglobin means that the relationship between how much oxygen haemoglobin binds, and the amount of oxygen available, is not a straight line but an initially steep and subsequently plateauing curve (Figure 9). The form of this curve means that each haemoglobin molecule loads its full complement of four oxygens in the lungs, where oxygen is

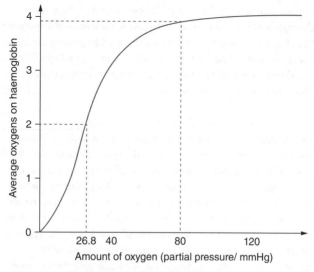

9. The oxygen-binding curve of haemoglobin.

plentiful, and every litre of blood contains the equivalent of 200 millilitres of oxygen. The very flat nature of the right-hand part of the haemoglobin curve in Figure 9 implies that humans will still be able to breathe in air with less oxygen per litre than in normal air at sea level and this is indeed the case; healthy people can travel thousands of metres up into mountains without experiencing breathing difficulties, although the very highest mountains in the world are high enough to create problems. In the tissues, where oxygen is being used so its concentration is lower, the haemoglobin curve (Figure 9) is at its steepest.

In a resting person, the oxygen concentration in the tissues drives haemoglobin molecules to release, on average, one of their oxygen molecules, which will diffuse to nearby cells. This leaves a significant reserve which can be called upon when tissue oxygen is more depleted, for example by strong exercise; here, the haemoglobin will give up more of its oxygen to meet demand. As it gives up oxygen, haemoglobin changes colour from red to a more bluish hue: this is why pale-skinned people suffering oxygen deprivation look noticeably blue. They might also look blue when cold ('cyanosis'); this is due to blood flow to the skin being restricted to conserve heat, leaving haemoglobin already in the skin to remain for longer and therefore give up more of its oxygen, over time, to meet the demands of the skin cells.

When cells 'burn' sugar in oxygen to make energy, they produce as byproducts water, heat, and carbon dioxide. Carbon dioxide and heat both encourage haemoglobin to offload oxygen molecules (the Bohr effect); together, these increase still further the ability of haemoglobin to deliver oxygen exactly where it is most demanded. When it enters the blood, carbon dioxide dissolves well and makes an equilibrium with carbonic acid and thence bicarbonate (Figure 10). In a healthy person, around 70 per cent of carbon dioxide produced is in the form of bicarbonate, with the rest being dissolved carbon dioxide and carbon dioxide bound to haemoglobin. Carbon dioxide does not compete to bind at the same place on haemoglobin that oxygens occupy but, nevertheless, carbon dioxide binds most easily when haemoglobin has already lost some of its oxygens (the Haldane effect). Thus, the more oxygen that is being released by the haemoglobin, the more carbon dioxide it can carry away—just what is needed in hard-working tissues. In the lungs, the blood passes close to the air, which has much oxygen but very little carbon dioxide. Here the Haldane effect is once again useful, as incoming oxygen encourages haemoglobin to let go of its carbon dioxide.

$$CO_2 + H_2O \rightleftharpoons H_2CO_3 \rightleftharpoons H^+ + HCO_3^-$$

| carbon dioxide | water | carbonic acid | hydrogen | bicarbonate ion |

10. The equilibria between carbon dioxide, carbonic acid, and bicarbonate.

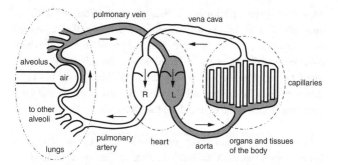

11. A diagrammatic representation of the circulation. The right side of the heart (shown on the left here as if looking at someone face-on) pumps blood into the lungs, where it flows through alveolar capillaries and collects oxygen (grey shading) from the air. The oxygenated blood flows via the pulmonary vein to the left side of the heart, which pumps it to the tissues of the body (including those of the heart itself) where oxygen is needed. This diagram is intended to represent the flow in the vascular system but makes no attempt to represent the complexities of real heart and vascular anatomy.

The working of this whole system depends on a constant flow of blood, to carry oxygen from the lungs to the tissues and to carry carbon dioxide away from the tissues and back to the lungs. This flow is achieved by having all of the blood moving in a one-way flow along closed blood vessels that together form a giant circuit around the body (Figure 11). Taking the lungs as an arbitrary starting place along this circuit, blood passes through the alveolar capillaries very quickly, spending only about one second there, because it is pushed out by incoming blood behind it. The capillaries join to form larger vessels, venules, which ultimately unite to form a very large vessel, the pulmonary (= 'pertaining to lungs') vein. The pulmonary vein carries all of the blood leaving the lungs to the left side of the heart. The heart is essentially a muscular bag equipped with one-way valves. When it relaxes, its inlet valves open and blood can enter; when it contracts, its inlet valves shut and its outlet valves open, forcing blood out at high pressure into the aorta and its branches. These arteries enter the tissues and branch, first into arterioles and then into fine capillaries that ramify through the tissues, bringing oxygenated blood close to the cells that need it. As in the lung, the capillaries unite to form venules, then veins, and these lead on to the vena cava, the great vein that collects blood returning from the body and feeds it to the right side of the heart. This right side of the heart, working in a similar manner to the left side, pumps blood into the pulmonary artery. This enters the lungs, splits up into arterioles, and these lead to the alveolar capillaries at which we started. There is therefore a continuous flow, driven by two synchronized pumps (the two sides of the heart); oxygenated blood flows from lungs to tissues and blood with reduced oxygen and more carbon dioxide flows back from tissues to lungs. The same blood flow carries glucose (and other nutrients) from the gut and liver to the tissues for use in generating energy.

Blood flow itself is not enough: with oxygen being removed from the air in the alveoli and carbon dioxide being added to it, that air would soon become too stale to support life. It is therefore necessary that old air is expelled and new air brought in, something achieved by the movements of breathing. Lungs are situated in a body cavity, the thorax, that is airtight except for the windpipe that carries air to the alveoli. The edge of the thorax is flexible; the front, back, and sides consist of a cage of ribs separated by a double row of intercostal muscles, while the lower boundary of the thorax is a domed sheet of muscle, the diaphragm. Normal breathing in is powered by an expansion of the volume of the thorax, driven by contraction of the diaphragm, which flattens it downwards, and by contraction of the outside row of intercostal muscles, which causes the ribcage to expand. For very deep breaths, other muscles can act to expand the ribcage even further. The lungs expand into the space available, being forced to do so by the inrush of air driven by imbalance between atmospheric pressure and the lack of pressure from the ribs to push against it. The inrushing air swirls into the airways of the lung, bringing new oxygen. Breathing out is powered by a combination of the diaphragm relaxing back upward into its dome shape, the inner row of intercostal muscles contracting the ribcage, and the lungs' own elasticity. This contraction expels air, and the cycle repeats.

Humans sometimes sit quietly, as when reading a book, and are sometimes very energetic indeed. Clearly the lungs and heart need to run at the correct level to deliver enough oxygen in both circumstances. Use of oxygen generates carbon dioxide and, by producing carbonic acid, excess carbon dioxide makes blood unusually acidic. This effect is used to sense the need to breathe more deeply to compensate for increased oxygen use. The control of breathing is complicated. Brain circuits set up a basic resting rhythm of one breath every five seconds or so. If the concentration of carbon dioxide in the body rises, it forms carbonic acid and body fluids become more acidic; this is especially so in the brain, the fluids of which have little capacity to buffer changes in acidity. Specialized parts of the brain detect this increasing acidity and signal to the breathing control centres to increase the rate and, especially, depth of breathing. This can mean a fifteenfold increase in the air volume breathed per minute. This flushes carbon dioxide out of the lungs and blood, and also of course brings more oxygen in, and breathing returns to normal when the excess acidity is corrected by loss of carbon dioxide. It is noticeable, though, that athletes show a transition to deep, powerful breathing without any evidence of abnormally high carbon dioxide. It is assumed that the brain initiates heavier breathing in anticipation of need as it initiates vigorous exercise. Breathing rate can be influenced by things other than carbon dioxide; sudden immersion in cold water can, for example, initiate gasping that has nothing to do with carbon dioxide. Also, we can consciously choose to pant or to hold our breath (up to a point; when carbon dioxide climbs high enough, the automatic systems in the brain override conscious breath-holding and force breathing to begin. This is why most people who drown have lungs full of water, but people whose body enters water after death, when these systems are no longer running, do not).

It will be noted that this system, working by monitoring carbon dioxide, does not sense oxygen itself. This is why oxygen-poor atmospheres (e.g. cryogenic facilities in which normal air might be displaced by nitrogen evaporating from tanks of liquid nitrogen) are so dangerous; as long as carbon dioxide levels are normal, the body gives no strong warning that there is no oxygen in the air and someone in that circumstance can simply lose consciousness and die, probably without ever knowing there is anything wrong. The advice sometimes given to free divers, to deliberately breathe hard and fast before diving to 'stock up on oxygen', is dangerous for the same reason. To begin with, 'stocking up on oxygen' is a nonsense—haemoglobin leaving the lungs will be almost completely saturated with oxygen in normal breathing and there is no significant extra storage available. What the hard breathing does do is remove too much carbon dioxide, allowing more time to elapse under water before carbon dioxide levels rise so much they force someone to surface and breathe. But, in this longer period, more oxygen will be used and there is a risk that the swimmer will simply lose consciousness and drown. There is some oxygen sensing in the body, but it responds only when available oxygen is very low, and the response is often too weak or late to save someone in an acute situation.

As breathing has to be matched to demand, so does the rate of circulation of blood. The heart has its own internal control systems, in which 'autorhythmic' cells constitute a pacemaker that, in the absence of other factors, send signals to the rest of the heart to drive contractions at around 100 beats per minute. External influences can slow this down (as when the body is at rest) or speed it up (as in vigorous exercise). The most important driver of increased heart pumping in response to demand comes from the nervous system. In exercise, the sympathetic nervous system signals to the pacemaker of the heart to increase its rate of firing, and to increase the power of muscle contraction in each stroke. The nervous system calms the heart down when demand has ended.

Heat

One important product of the body's energy budget is temperature control. Active tissues produce heat. When the body is at rest, internal organs such as the liver and brain dominate heat production, but in exercise muscles can produce around twenty times as much heat as the rest of the body. Humans, like all mammals, need to maintain their core temperature between about 35.5 and 37.8 °C, whether they live in the Arctic or tropics, and whether they are sprinting or sleeping. Thus, under some circumstances, the problem is generating and retaining enough heat while, under others, the problem is disposing of excess heat.

One major mechanism for heat control again uses the blood system as a transporter. If the body is too warm, blood vessels in the skin dilate to admit a strong blood flow, so that hot blood from the body's core enters the skin to lose heat to it. This is why hot people often look red, at least if their skin is pale enough for effects like that to show. At the same time, the sweat glands of the skin cover it in moisture. Liquids are always a mixture of fast-moving and slower-moving molecules, and the temperature of the liquid is a function of the proportion of molecules moving at different speeds. Where liquid meets air, faster moving molecules of the liquid break free and enter the air more easily than the slower ones, so the molecules left behind now have a lower average speed and therefore a lower temperature. For this reason, sweating to make the skin damp cools it, cools the blood flowing through it, and cools the body. In extreme conditions, the body can produce two litres of perspiration every hour, which is why vigorous exercise in hot weather results in thirst. The skin also cools by radiating heat and directly shedding the heat to cooler air. If even this is not enough to cool the body, behavioural instincts drive the seeking of shade or cool water, cessation of strenuous activity, and, in the case of clothed humans, to shed some or all of that clothing.

If the body is too cool, it conserves heat by constricting blood vessels in the skin, reducing flow of blood through the cool skin, and therefore reducing heat loss. This reduced flow of blood in the skin is why cold pale-skinned people often look white or very pale blue. Sweating almost stops, but metabolic rate increases, especially in a type of fat cell called 'brown fat'; brown fat cells can 'burn' fat without generating ATP, so that all of the energy goes to heat rather than being stored in high-energy chemicals. If these mechanisms are inadequate, shivering begins: shivering is a way of working muscles to generate heat without generating overall motion. This is often accompanied by postural changes to minimize exposed area. There are also behavioural instincts to seek shelter and, in the case of humans, to put on more clothes, to eat and drink warm foods, and to light a fire.

These mechanisms and behaviours are mostly under the control of the hypothalamus, a region of the brain that receives temperature information from many sites in the body. The hypothalamus senses departures from normal temperature and initiates the appropriate mechanisms to restore it. This is an example of an idea very important in physiology: homeostasis, or 'staying the same'. It is through having a large number of homeostatic mechanisms that we can keep our bodies functioning within tight limits even in the face of very different environmental and behavioural influences. Homeostasis will be considered in more detail in Chapter 3.

Energy balance

So far, we have considered keeping the energy budget with respect to breathing (carbon dioxide concentration) and temperature, but not with respect to the input of food. Because food can be stored, it does not require the fast-acting homeostatic mechanisms that are needed for breathing, but it is still important that, long-term, the amount of food-derived energy taken into the body balances that which is used. Consistent imbalance between energy input and use results in starvation or obesity. It is noticeable, though, that for most people the balance is kept remarkably well. With the exception of people who indulge in periodic dieting, most people's body weight is stable even over many years. It may not be stable at the weight someone would most like it to be, but it is stable. This argues for a strong balancing mechanism, and work over recent decades has begun to uncover at least some (but far from all) of its details.

It is clear that, when food is being eaten, the digestive tract signals, via nerves, information about what is inside it, and that may be one element of satiety. Rising blood glucose is recognized in the brain, and also drives satiety. Fat cells signal their own reserves by secreting the hormone leptin, which suppresses appetite. People with faulty leptin-signalling systems are very hungry and tend to become obese through serious overeating. In addition to these feedback systems, there are also feed-forward influences: cold weather, for example, tends to increase appetite for energy-rich foods even before keeping warm has significantly depleted fat stores.

Waste

The glucose-burning energy production system that has been the focus of this chapter has two waste products, carbon dioxide which is breathed out, and water. Depending on the demands of body cooling, this water may be lost as perspiration or need to be expelled another way (lest the body become swollen and diluted with it). Other metabolic processes in the body create other waste products. Amino acids obtained by the digestion of proteins, for example, can either be used directly for the construction of new body proteins (see Chapter 7) or used for energy. Use of amino acids for energy generates a chemical waste product, urea, which needs to be disposed of. Many other minor metabolic pathways, including those that detoxify dangerous components of foods, or process drugs, generate their own wastes. All of these wastes are disposed of in a concentrated solution, urine, produced in the kidneys and stored in the bladder before release.

In principle, one could imagine a kidney that consisted of specific export mechanisms, each of which recognized one type of waste product in the blood and expelled it to a urine space. Kidneys do indeed contain some such systems, but an organ that worked in entirely this way would be a bad idea because there would be no way for it to deal with an 'unexpected' waste product, for example a metabolite of a component of an exotic fruit that humans did not evolve to eat. Therefore, the kidney uses a different but more flexible solution. Within the organ, blood capillaries form tight 'knots' of leaky vessels—glomeruli—that are surrounded by cells that form a sheet with very tiny holes. These holes are so small that they will permit passage of water, salts, and small molecules such as urea, but will not allow large proteins to pass easily. Water, salts, and small molecules therefore pass through the filter and into the inside of a tube—the nephron—beyond (Figure 12).

As the fluid progresses along the nephron, it passes the cells of the nephron walls. These cells have specialized transport mechanisms to recover useful molecules, such as glucose and salts, that the body cannot afford to lose. The cells therefore recover these molecules, at the cost of considerable energy expenditure, leaving

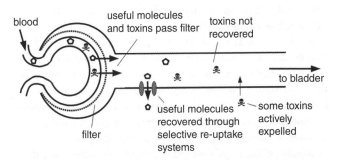

12. **An idealized kidney nephron. Small molecules, whether useful or waste, are filtered from blood into the nephron. Here useful molecules are actively recovered while toxins remain in the urine to be expelled. Some toxins are also actively expelled in the nephron, independently of the filter. In this diagram, anatomy has been greatly simplified.**

the unwanted molecules such as urea in the remaining liquid. Much of the water is also recovered, in a system that can be regulated by hormones so that the body can control how much water it loses (see Chapter 3).

Normal kidneys contain around a million nephrons each, and about a fifth of the blood that leaves the heart passes through them. They are therefore very well placed to make rapid adjustments to blood chemistry, and they contain many systems that are responsive to measurements of blood, such as the concentration of different salts, and can adjust how much of each component is recovered from, or lost to, urine. In some conditions, though, these systems are overwhelmed. This happens in uncontrolled diabetes mellitus, in which body cells become unable to take up glucose from the blood properly and the blood concentration of glucose is much too high. This overwhelms the glucose recovery systems of the kidney, leaving a significant amount of glucose in the urine. Osmotic effects of this interfere with water recovery too, so that too much water is lost, hence the name of the disease: 'diabetes' means too much urine flow, and 'mellitus' means sweet-tasting.

This chapter, focusing on energy as just one aspect of body function, has introduced many different organs (mouth, oesophagus, stomach, intestines, liver, lungs, heart, kidneys), body systems that serve many organs at the same time (the blood circulation and the nervous system), and a variety of mechanisms that have to cooperate for energy use to work. Beginning with a focus on almost any other major aspect of life would have involved a similar number of organs and systems, because human life depends on the smooth coordination of many different mechanisms. The regulation and coordination of activity is critical to physiology, and will be the topic of Chapter 3.

Chapter 3
Homeostasis: the stability of the internal environment

The need for stability

The biochemical reactions of life, of which the energy-handling metabolic reactions described in Chapter 2 constitute just a subset, take place properly only within a tightly defined range of physical and chemical parameters. If saltiness, acidity, concentration of nutrients and wastes, or temperature are allowed to drift too far from their optimum, at least some reactions run either too vigorously or too slowly, and life becomes unsustainable. The body therefore has to maintain its internal conditions within narrow limits. In doing this, it faces two problems. One is that the conditions of its surroundings change, from the heat of a sunny day to the chill of a starlit night, or from the dryness of desert dunes to the humidity of a tropical rainforest. The other problem is that the activities of the body itself change as it exercises, rests, eats, becomes pregnant, etc. That the maintainence of a stable internal environment is itself an important, active physiological process was recognized by Claude Bernard in 1865. He understood that the body must have mechanisms to measure its internal state and, if the measurements suggest a drift from ideal conditions, to invoke a response that will oppose that drift. Later, the physiologist Walter Cannon gave it the name 'homeostasis', from the Greek *homoios* (similar) and *stasis* (standing still).

Elementary cybernetics

Control systems, whether natural or artificial, can be divided into two broad classes: open-loop and closed-loop. Open-loop systems (Figure 13(a)) consist of a device under a control system that pays no attention to how well the device is working. A homely example is a cheap electric room heater, which has a basic power control for how much electricity it turns into heat (e.g. a knob that selects how many radiant bars are active). The user selects the power setting, but the temperature achieved in a given room will depend not only on this power but on the outside temperature, on whether the window is open or closed, on whether sunlight is streaming in, etc. It is a terrible method for achieving a predictable temperature, and heaters like this are useful only when people are willing to switch their power up or down according to how cold they feel. Closed-loop devices (Figure 13(b)) have inbuilt control systems to do what the users of open-loop devices must do for themselves. They sample what has been achieved by the device, compare it to a desired outcome, and alter the power of the device accordingly. An example would be a room heater equipped with a thermostat. In this case, the user selects the desired room temperature, and the thermostat adjusts the power of the heater according to the actual temperature of the room. Within limits, the device will hold the room temperature steady on a hot or a cold day and whether or not the window is open.

Of these types of control, the closed-loop type is clearly best suited to homeostasis, and it is not surprising that it is widespread in physiology. It has been intensively studied, as a branch of physiology and also as a science in its own right; in 1948, Norbert Wiener named this science cybernetics, from the Greek for 'steerer', defining it as 'the scientific study of control...in the animal and the machine'. Closed-loop devices are characterized by the presence of a channel of communication that feeds the result

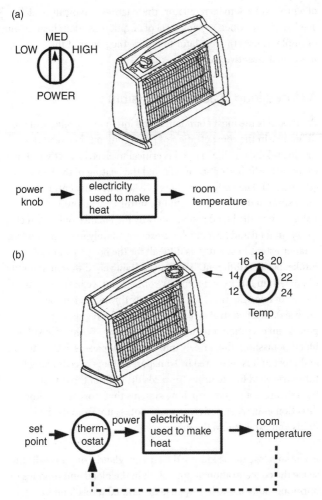

13. **Open- and closed-loop control, illustrated with a simple domestic example: (a) depicts open-loop control of an electric heater, in which the control knob selects the power with which the heater will operate, and it will do this however hot the room; (b) depicts closed-loop control, in which the control knob selects a desired temperature and the heater controls its own power according to how warm the room actually is. In the schematic diagram, the feedback loop is shown by the dotted line.**

of a process back to its regulator; this channel is typically called 'feedback' and, when a rising level of output (e.g. room heat, in our example) causes the regulator to reduce the power of the process, it is called 'negative feedback'.

A closed loop in action: potassium

Potassium is the most abundant metal ion in living cells, and the difference in the concentration of potassium ions between the inside and the outside of a cell is critical to electrical activity, for example the 'firing' of nerve cells and the beating of the heart (Chapter 5). The body must therefore keep the concentration of potassium in its fluids under tight control. The amount of potassium in the blood is set mainly by the balance of how much we eat in food and how much we excrete, mainly in urine and to a smaller extent in sweat. It is difficult for the body to adjust its intake, because different foods can contain very different amounts of potassium for the same energy content and we have no sensory means of knowing this (we know it only through scientific analysis). Most control is therefore exerted by regulating potassium excretion into urine by the kidneys. Within the kidney, blood potassium filters freely into the urine space (see Chapter 2), and most of it is recovered by transport systems in cells along the tubes travelled by the urine. Towards the ends of those tubes, though, are potassium-expelling systems that work in the other direction and, when active, can take potassium from the body fluids and expel it into the urine.

An organ close to the kidney, the adrenal gland, contains cells that sense the concentration of potassium in the blood and produce the hormone aldosterone in response to elevated levels (and in response to other things, but potassium is a particularly powerful influence). Aldosterone activates the potassium-expelling transport systems of the kidney, and so acts to reduce the

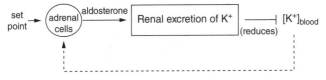

14. The closed-loop system that regulates the concentration of potassium ('[K$^+$]') in the blood. The internal contents of the rectangular box are complicated, but can be lumped together under a description of their overall effect. This 'black-box' treatment of some mechanisms is common in diagrams of physiology, and allows a reader interested in the overall process not to get lost in detail.

concentration of the ion in the blood, restoring it to its set-point (Figure 14).

Integral and derivative control

The simplest kind of closed-loop system is one that uses proportional control: the larger the difference between the current state (e.g. temperature) and the desired values, the harder the regulator makes the device work to reduce that difference towards zero. This makes intuitive sense, and seems to be the description of feedback most commonly given in physiology textbooks, but it suffers from a serious flaw: the closer the state of the system is to the desired state, the less powerfully the device works and it would take a literal eternity to restore the state to its exact intended level after a large perturbation (Figure 15(a)). An alternative is integral control, in which the regulator responds not to the current difference between the actual state and intended state, but to the integral of that difference over recent time; within limits, the longer a difference has persisted, the more powerful is the correction given (Figure 15(b)). This has the advantage that it will restore a perturbed system to its set-point, but the disadvantage that, because it is sensitive to time as well as extent of error, integration causes a delay in responding to a very rapid-onset error. Also, because integral

(a)

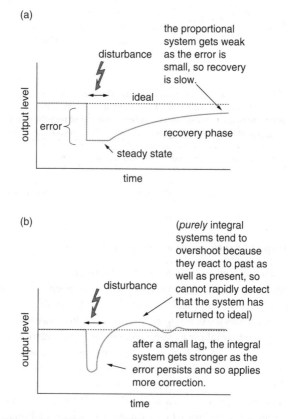

the proportional system gets weak as the error is small, so recovery is slow.

disturbance

ideal

output level

error

recovery phase

steady state

time

(b)

(*purely* integral systems tend to overshoot because they react to past as well as present, so cannot rapidly detect that the system has returned to ideal)

disturbance

output level

after a small lag, the integral system gets stronger as the error persists and so applies more correction.

time

15. The ways in which proportional control (a) and integral control (b) respond to a disturbance. Most biological homeostatic systems are a combination of both.

control responds to the total effect of error over time, the restorative signal is not switched off as soon as the ideal system state is reached, and it is common for this type of control to overshoot (Figure 15(b)). Useful systems therefore use a combination of proportional and integral control, and the best also add a third element, derivative control, which boosts power of the restoring system in proportion to the rate at which the system's state fell from the ideal.

Calcium homeostasis

Calcium ions have a large range of physiological functions: they are used structurally in the skeleton and to assist cells' mutual adhesion, used as signals inside a range of cells, and are needed for defensive processes such as blood clotting. The blood concentration of calcium ions has to be maintained within a narrow range, even in the face of major perturbations, such as a nursing mother excreting large amounts of calcium into her milk (Chapter 8). The processes that contribute to maintaining calcium concentration include intestinal uptake of calcium, excretion of calcium by the kidneys, and uptake or release of calcium from bone. Calcium levels are sensed by cells in the parathyroid gland in the neck, and these cells secrete parathyroid hormone (PTH) when blood calcium levels are too low. PTH causes bone to give up its calcium and kidneys to recover more calcium from urine (Figure 16(a)).

What has been described so far is proportional control, and detailed mathematical modelling shows that it cannot account for how well mammals cope with sudden large demands on their stores of calcium; on its own, proportional control would restore blood calcium concentration at a steady but too-low level (Figure 16(b)). PTH, however, also causes kidney cells to produce another hormone, 1,25-DHCC, from vitamin D. This stimulates the intestine to absorb more calcium from food. Because the production of 1,25-DHCC is proportional to PTH concentration, the total amount of accumulating 1,25-DHCC is proportional to how much PTH is present and for how long it has been present (that is, to the integral of PTH concentration with time). This means that, while the contribution of bones and kidneys to calcium homeostasis is under PTH-mediated proportional control, the contribution of intestines is under 1,25-DHCC-mediated integral control (Figure 16(c)). Computer modelling of this complete system demonstrates that it can produce the robust

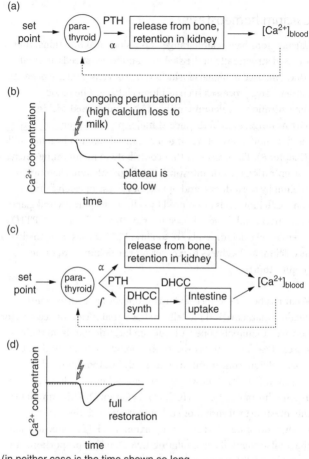

(a)

set point → para-thyroid —PTH, α→ release from bone, retention in kidney → $[Ca^{2+}]_{blood}$

(b)

Ca^{2+} concentration vs time

ongoing perturbation (high calcium loss to milk)

plateau is too low

(c)

set point → para-thyroid, α and ∫ PTH, DHCC → release from bone, retention in kidney; DHCC synth → Intestine uptake → $[Ca^{2+}]_{blood}$

(d)

Ca^{2+} concentration vs time

full restoration

(in neither case is the time shown so long that bone reserves become exhausted)

16. Calcium homeostasis: (a) shows the PTH proportional system only, which fails to restore blood calcium levels ($[Ca^{2+}]_{blood}$) after a sudden drop (b); (c) shows the combined proportional and integral system, which restores blood calcium levels fully (d).

correction of blood calcium concentration following sudden perturbation (Figure 16(d)).

Controlling blood sugar

One of the most widely known of the body's homeostatic systems, because it is relevant to a common and serious disease, controls the concentration of glucose in the blood. As noted in Chapter 2, the concentration of glucose in the blood can be altered by many activities. Digesting carbohydrates in food, and turning stored glycogen into glucose for release into the blood, act to increase blood glucose concentration. Using glucose to generate ATP, or using it to build new molecules of glycogen, reduces blood glucose concentration. Obviously, the body needs to maintain an adequate reserve of glucose in the blood to provide for the energy needs of tissues but, less obviously, it also needs to prevent the concentration of glucose becoming too high because the molecule is somewhat reactive and can damage blood vessel walls and cells of the nervous system if present in excess. The control mechanisms that maintain glucose homeostasis have to be able to react relatively quickly, because blood glucose can change rapidly, either in response to strenuous exercise or to intake of sugary food or drink.

Glucose control is centred mainly on specialized regions of the pancreas called pancreatic islets. Within the islets are two types of endocrine (hormone-secreting) cell, alpha cells and beta cells, which work in a harmonious opposition to keep blood glucose levels well controlled. Beta cells secrete two hormones, insulin and amylin, in proportion to the concentration of glucose in the blood. Insulin has several effects in the body, one of which is to induce cells to take up glucose from the fluid that surrounds them, and thus indirectly from blood, and to use that glucose either for local ATP generation or, if they already have enough ATP, for storage as

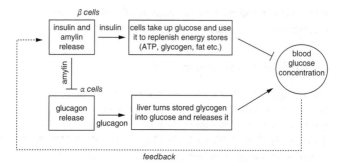

17. The main pathways that control concentrations of glucose in the blood. In this diagram, as in all others in this book, the arrow represents activation or increase, and the 'T' shape represents inhibition or decrease.

glycogen or fat. Insulin also inhibits the production of glucose from cellular energy stores such as glycogen, protein, or fat. Its general effect is therefore to remove glucose from the circulation and to prevent more being put there (Figure 17, top row).

Alpha cells, on the other hand, make the hormone glucagon and will do so all the time unless they are inhibited from doing so by amylin released from the beta cells. If the beta cells are secreting a large amount of insulin and amylin, which will be the case if blood glucose is high, glucagon production is deeply inhibited. When blood glucose levels fall, the beta cells make very little insulin and amylin, so glucagon production by the alpha cells can proceed unimpeded. Glucagon acts mainly on the liver, and causes its cells to produce glucose from breakdown of glycogen or to make it from other small molecules (gluconeogenesis), restoring the levels in the blood (Figure 17, bottom row). The push-and-pull natures of this feedback system, with the pathways in the top and bottom rows of Figure 17 each playing an active role, allow it to respond rapidly to errors in either direction. In the normal state of the body, both pathways are somewhat active and the overall concentration of glucose in the blood is set by a balance of their activities. There are, though, complexities. For example, insulin levels after a meal are not as steady as Figure 17 would suggest they should be, but actually come in pulses every five minutes or so. These pulses may be important in preventing responding cells becoming insensitive to insulin ('adaptation': see Chapter 4). There is still much about this system we do not understand.

Some people have an immune disease that destroys their beta cells and, as a consequence, they cannot make enough insulin or amylin. Their general body cells therefore receive no signal telling them to take up glucose from the blood, and the cells are in a near starving state and have to use their own reserves of glycogen and fat for energy. The blood, on the other hand, has far too much glucose, partly because it has not been taken up, and partly because lack of amylin allows glucagon to be produced and to drive blood glucose concentrations up even further, at least as long as glycogen stores in the liver are capable of delivering it. This high concentration of glucose causes damage to blood vessel walls, and also disturbs kidney function as explained in Chapter 2. The result is someone who produces too much urine, is therefore very thirsty, is very hungry, and may eat a great deal but still lose weight as the glucose they eat cannot reach starving cells. In extreme cases, he may suffer severe metabolic imbalances and even lapse into a coma, possibly terminal. This disease, diabetes mellitus type I, is treated by regular injections of insulin.

People with diabetes mellitus type II, on the other hand, can have normal insulin production but their body cells become insensitive to it; the reasons behind this insensitivity are complex and still being worked out. It is clear that there is a strong genetic component to risk. There is also a link to obesity, fat cells producing a molecule that depresses the production of accessory proteins needed for insulin-triggered glucose uptake, and this effect seems to be strongest in sedentary people. For this reason, people at genetic risk of type II diabetes are strongly advised to manage their weight and to exercise (as are we all, but it is especially important for people with type II diabetes). While insulin and glucagon are of particularly great importance to glucose homeostasis, they are not the only influences, and the system is also somewhat sensitive to other hormones from various parts of the body including the digestive system and brain. Some of these are attracting intense interest as possible drug targets for better management of type II diabetes.

Light relief

One of the simplest homeostatic mechanisms to observe, in oneself or someone else, is the adjustment of the pupil of the eye. Eyes, discussed in more detail in Chapter 4, contain cells that translate light into the firing of nerve cells, and thus provide the brain with an encoded representation of the scene in front of the eye. The light-sensing cells have a dynamic range, from what they see as white to what they see as black, of about one hundred to one, which is at least a hundred times less than the total range of light intensities, from sunny seaside to moonlit woods, in which humans can use their sense of vision. The eye has several strategies for managing to work across a much larger range of brightnesses than any one of its own cells can use. One strategy is to have different cells for different brightnesses (see Chapter 4), and another is to optimize the eye for relatively dim daylight and to shut out excessive light when the day is bright. This shutting out is done by constriction of the pupil, the transparent 'hole' through which light enters the eye. In dim light, the pupil of a young adult is about 9 millimetres across while in bright light it can close down to 2–3 millimetres, cutting down the amount of light entering the eye by about tenfold (ten, because the area of a circle scales with the square of its diameter).

Vision is a fast sense, fast enough for a human running over rough ground to detect hazards before falling over them, and fast enough to enable hunting of prey and usually fast enough to avoid becoming the prey of some larger predator. Adaptation of the pupil as someone runs from a shaded wood into bright grassland therefore has to be rapid—much more rapid than could be achieved by the ponderous secretion and circulation of hormones that has been the basis of the homeostatic systems we have considered so far. In general, whenever the body has to do something quickly, it uses the nervous system, a network of 'living wires' that conduct electrical signals at high speed from precise place to precise place (see Chapter 5). Like the glucose homeostat, the light homeostat uses push-and-pull control.

The pupil itself is surrounded by the iris, which has a set of muscle fibres that run circumferentially around the pupil, and a set of muscle fibres that run radially from it, like spokes from a bicycle hub (Figure 18). When too much light enters the eye, light-sensitive cells in the eye (photosensitive ganglion cells) send nerve impulses to the brain (Chapter 5) and, in response, the brain sends signals along a nerve to a local information-processing centre, the ciliary ganglion. The ganglion initiates the firing of nerves that activate the circumferential muscles of the iris, pulling the pupil closed. At the same time it stops sending nerve pulses to stimulate the radial muscles that would tend to open the pupil. In dim light, the nerves controlling the radial muscles activate, actively pulling the pupil open again. This effect can be observed easily in a friend asked to look at a computer screen in a dimly lit room, while the screen brightness is turned down and up. As with the other homeostatic mechanisms described in this chapter, the core homeostat is not the only influence on the output. Even in steady light levels, someone's pupils will tend to contract if they are bored or confused and will tend to expand if they are interested in something or in someone, particularly if that interest is romantic. Many recreational drugs also affect the pupil, opioids such as diamorphine ('heroin') causing it to constrict and LSD, MDMA ('ecstasy'), and cocaine causing it to dilate.

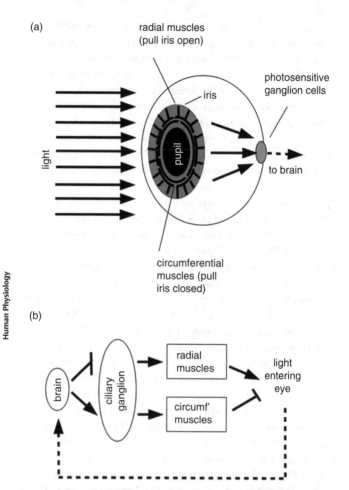

18. The homeostat for illumination in the eye: (a) a simplified version of the anatomy of the system; (b) the homeostatic system schematically.

From needs to drives

A very large number of homeostatic responses involve the nervous system, including the control of respiration and heartbeat described in Chapter 2. While some of these work using relatively simple effectors (the muscles that control the iris and pupil, or that control breathing) and operate below the level of consciousness, others use very sophisticated mechanisms in the brain, including reasoning and problem-solving. They therefore provide a link between physiology and psychology. Detection of dehydration, for example, evokes unconscious responses, such as more powerful water recovery by the kidney, and also the conscious feeling of thirst. Thirst is an example of a 'drive'—a desire for a particular behaviour produced by departure from a homeostatic ideal. If a glass of water happens to be at hand, the drive can be satisfied easily with barely a thought but, under other circumstances, the drive may co-opt the most powerful aspects of conscious brain function. A thirsty hiker may, for example, consult a map to find nearby villages, interpret the size and features of each one to make an intelligent guess about which might contain a shop or pub, compute a navigational course to the most promising, and, having found one, use complex spoken language to conduct the drink-buying transaction.

Many drives, such as desires for water, food, warmth, sleep, or a lavatory, can be linked very clearly to individual physiological homeostasis. Others, such as lust, though clearly linked to aspects of human physiology, are less obviously linked to homeostasis of an individual (even if they link to maintenance of a population). Drives are not just effectors of homeostasis, therefore, and it may be more correct to view them as something belonging to the physiology of brain and behaviour that happens, in some cases, to have been co-opted by homeostatic systems where this works. Drives and behaviour will be discussed further in Chapter 5.

It must be stressed that all of the homeostatic systems described in this chapter have been presented in a simplified form, with only the most important regulators mentioned. This has been done to highlight as clearly as possible some important ideas behind human physiology, without burying the most important information in vast numbers of additional influences and effects. The minor components ignored here may nevertheless be important for medicine, and many are used as drug targets for making small, safe differences to the way a particular system works, to help control a disease.

Chapter 4
Sensation

Bringing the outside in

Sensation is a catch-all term for monitoring any state and feeding it into a physiological process. So far, we have focused on the sensation of internal states, such as chemical concentration. Though physiologically critical, this type of sensation is largely hidden from everyday human experience, and when people talk of their 'senses' they usually mean the five senses by which they *consciously* monitor features of the outside world. These senses—vision, hearing, smell, taste, and touch—provide rich flows of information and most make use of specialized organs. In all five cases, the sensory system combines two functions: measurement of a stimulus, and encoding it in a way that can be transmitted via a nerve into the brain. In addition, the brain may signal back to the sensing system to modulate the way that it works; right now, your brain is sending information to your eyes to change the precise direction of their gaze, as you read these sentences from left to right.

In trying to understand organs of sensation, it helps to forget, for a moment, the complexity of the human mind and its will and consciousness, and consider our much simpler ancestors in which these sense organs first evolved. Simple animals show a fairly tight coupling between stimulus and response (something pokes them, they move away). What these animals needed of their sensory systems was rapid and accurate detection of an environmental change *relevant* to driving a behavioural response. This is a quite different requirement from providing a faithful and unbiased report of all external conditions, and we therefore should not be surprised that their sensory systems evolved to provide highly selective information strongly biased to detecting particular environmental features and changes. In the long evolutionary path from such animals to ourselves, these sensory systems seem to have changed very little; what has (probably) changed is the brain's ability to construct detailed models of the world from biased and selective data.

Touch

Touch is one of the most ancient senses, and is the one that is most widely distributed across the body. Touch is really detection of mechanical stimulation, and there are several different types of mechanical stimulation, such as pressure and shear. There are also different concentrations of force, from the diffuse pressure of the seat probably underneath you now, to the highly focused pressure you would feel if you were to bite your lip. Subjectively, we are aware that we feel different types of touch differently; nobody would mistake a pin-prick for a gentle caress. The skin has several different types of touch sensor, each responding to different kinds of stimulation, but it seems that there is no simple relationship of the kind 'sensor 1 detects pins, sensor 2 detects water splashes' and so on. Rather, information from different types of sensor is integrated by the brain to produce the subjective experience. This book is much too short to consider all touch-sensing systems so, for the sake of understanding principles, we will consider just two that are particularly important in everyday life, and simply note the existence of many more.

The skin is a two-layered structure; the top, visible, layer is called the epidermis, and its main function is to form a seal at the edge

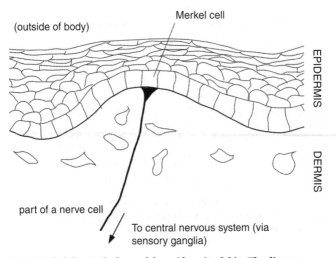

19. A Merkel disc in the base of the epidermis of skin. The diagram depicts a cross-section of a ridge at a fingertip (these ridges are those that show up as fingerprints).

(outside of body)

Merkel cell

EPIDERMIS

DERMIS

part of a nerve cell

To central nervous system (via sensory ganglia)

of the body (Chapter 7). Its cells are attached to one another strongly with cell-to-cell junctions. Below the epidermis, and not seen except in a graze injury, is the dermis, a layer of thick connective tissue that supports the epidermis. In the bottom layer of the epidermis can be found sensory structures called Merkel discs; they are arranged densely in skin of the fingertips and genitalia, both of which are sources of detailed touch information, and less densely in most other areas of the body. Anatomically, Merkel discs consist of Merkel cells clustered around the end of a nerve cell (Figure 19). Merkel cells have several unusual properties. One is that they are full of storage vesicles (tiny membrane-bound spaces) containing the neurotransmitter serotonin. Another is that their membranes contain a force-sensitive membrane protein, Piezo2. Piezo2 is normally closed to the passage of positive ions but it opens when under mechanical load. Because they stick strongly to their neighbouring epidermal cells, if the

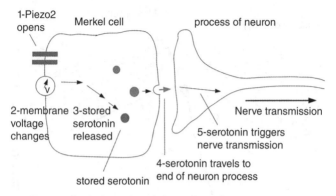

20. How mechanical opening of Piezo2 channels results in nerve signals being sent from a Merkel disc in the skin; the '1', '2', '3' refer to order of events.

skin experiences a mechanical force, the Merkel cells will experience it too.

Cells normally have a voltage across their membrane (see Chapter 5) due to unequal concentrations of certain ions inside and outside the cell maintained by ATP-driven pumps. When the Merkel cells' Piezo2 channels open due to mechanical force on the skin, the flow of ions through these channels alters the membrane voltage and initiates a series of biochemical events that culminate in the Merkel cells releasing some of their stored serotonin. The nearby nerve endings are responsive to serotonin, and initiate nerve transmission when they detect it, conveying news of the mechanical force to the brain (Figure 20). It takes very little force for all this to happen; the lightest touch of a feather is enough. The importance of Merkel discs in sensation of light touch is supported by the demonstration that, if the Merkel discs of mice are rendered inoperative by deliberate genetic mutation, the mice become less sensitive to touch on their paws. They are not, however, completely insensitive, showing that the various other types of touch sensor are also important.

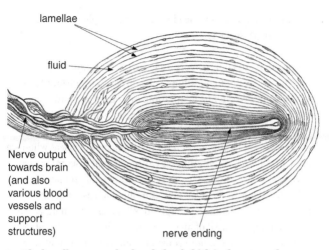

lamellae

fluid

Nerve output towards brain (and also various blood vessels and support structures)

nerve ending

21. The lamellar corpuscle, the whole of which is about 1 mm long.

Also in the skin, especially in areas such as fingertips, is another type of mechanical receptor called a lamellar corpuscle. It consists of a sensitive nerve ending encased in layer-upon-layer of fibrous 'shells', the lamellae of its name, each full of viscous fluid and with occasional elastic contacts spanning the fluid (Figure 21). If the skin above a lamellar corpuscle is suddenly pushed inwards, the corpuscle is distorted and the nerve ending is deformed, allowing more ions to flow across it and alter its voltage and, if this effect is large enough, causing the nerve ending to fire and send an impulse towards the brain. In response to the distortion, the fluid in the lamellae moves, which tends to relieve the mechanical force on the nerve ending and therefore the firing ceases. When the external cause of the skin distortion goes away, the now off-centre distribution of fluid in the lamellae causes its own distortion and stimulates the nerve fibre until the fluid returns to an even distribution. The system is therefore set up to be sensitive to changes in the position of the surface of the skin, rather than to keep reporting on an unchanging position, and the corpuscles are activated particularly well by rapid variations in the position of the skin surface as experienced when, for example, a fingertip strokes a rough surface.

The sensitivity of lamellar corpuscles to change, and their inactivity in constant conditions, introduces an important feature of many sensory systems: adaptation. Systems that show adaptation report changes but are much less stimulated by constancy. Adaptation is familiar from everyday experience; we feel the touch of clothing when we put it on, and of a seat when we sit down, yet, unless something is causing actual discomfort, we cease to be aware of these touch-type stimuli while remaining perfectly capable of feeling a new stimulus such as an insect landing on us. The same can apply to other senses; generally, we cease noticing the smell of an environment that was noticeable when we arrived, and ignore constant sounds to the extent that most long-haul passengers are capable of sleeping only a few metres from a roaring jet engine. Some adaptations take place at receptors themselves, as in the example above. Many take place instead in the neural circuitry to which sensors are connected, but all make sense in the context of a simple animal that needs to respond to changes. The natural focus of our senses on changes rather than the constant drives home the point that our information about the external world is partial and highly mediated.

Hearing

Sound is an essentially mechanical phenomenon: waves of alternating high and low pressure propagating through the air at, by definition, the speed of sound. Detecting sound therefore, like touch, requires a physiological system that can translate pressure waves into nerve signals. Low frequency sounds can be detected by the lamellar corpuscles described above; the presence of these corpuscles in the pancreas, in tissues associated with the intestine, and in the genital skin may well account for the fact that people often 'feel' strong, very deep sounds more than 'hear' them, as a vibration in the abdomen and, in men, the penis. Most sounds, though, are detected only by the ears.

Although the sounds humans hear travel by air, the actual detection of sound waves is done in a fluid-filled space, the cochlea. In life, the cochlea is wound into a tight spiral (Figure 22(a)), but Figure 22(b) depicts it as if it has been unwound. It is a conical tube full of a fluid called perilymph. For most of its length, the tube is divided transversely by another tube of triangular cross-section, filled with another fluid, endolymph. The endolymph tube is closed off at both ends. The broad ends of the perilymph spaces are also closed, but with thin flexible membranes called the oval window and the round window. By means that will be considered in a few paragraphs' time, sound waves enter the oval window into one perilymph space and, having crossed the endolymph space, the sound waves emerge from the round window as 'waste'. It is while they are crossing the endolymph space that the sound waves are detected. Within the endolymph spaces are delicate sensory cells, called hair cells (because of their shape, not because they have anything to do with actual hair). The cells are anchored to the basilar membrane that separates endolymph and perilymph, and they project fine processes called stereocilia. Stereocilia make fine contacts with one another and, in the case of the longer ones, with a delicate membrane above them (Figure 22(c)–(d)). If the basilar membrane is deflected up and down by sound waves, the stereocilia are forced to bend, pulling on the connections between them. This pulling opens ion channels and initiates nerve firing. An interesting feature of the cochlea is that sounds of different pitches travel across the endolymph tube in different places, high-pitched sounds tending to cross close to the wide end of the cochlea, low-pitched sounds tending to cross nearer its end, and intermediate pitches crossing at intermediate distances. The reasons for this are complex and are to do not only with the simple physics of the cochlea and its membranes but also active responses by cells themselves which sharpen the accuracy with which

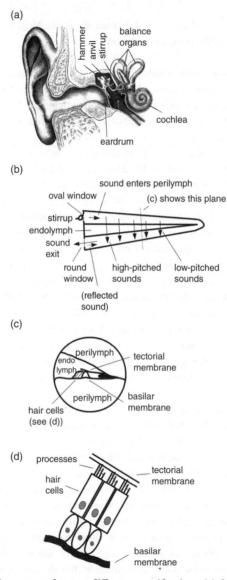

22. The human ear, drawn at different magnifications: (a) shows an anatomically realistic view, (b) shows the cochlea straightened out, (c) shows a cross-section of the cochlea, and (d) a detail of the hair cells.

different frequencies map to different parts of the basilar membrane. Receiving information from nerves activated by the hair cells, our brains interpret the information of which cells are active as pitch, of a single tone, or as the quality of a sound and probable identity of its source, for mixtures of frequencies that would be present in, for example, the quack of a duck.

The cochlea of the inner ear is, as has been mentioned, filled with fluid so sound has to be passed from the air to this fluid. This is problematic, because of an impedance mismatch. Impedance is a concept familiar from everyday life. Consider a bicycle with many gears. For any given rider on a flat surface, there is a gear for which the 'stiffness' of the pedals is a good match for the 'push' of the legs, and pedalling is efficient. In too low a gear, there is no opportunity to push properly because the pedal moves down too quickly, while in too high a gear the pedal is too hard to push and pushing tends to lift the rider up from the saddle instead. This hardness to push is impedance, and the gears are fitted to the bike to allow the impedance of the bike to be matched to an impedance that matches that of the rider's legs, so that power can be transferred efficiently. The fluid of the cochlea has a much higher impedance than air, so the ear needs an auditory equivalent of bicycle gears to transform impedance from that of air to that of fluid. This impedance matching is done by three small bones, the hammer, anvil, and stirrup (Figure 22(a)), and by the eardrum that is directly vibrated by air. The bones use a kind of lever action to turn a gentle force and large movement at the eardrum into a very forceful but small movement at the oval window of the cochlea, achieving the required impedance transformation. The eardrum itself receives sound directly from the air, via a tube from the pinna, the visible external ear. The pinna gathers sound and its complex shape adds small distortions to high frequency components. These distortions vary with the direction from which the sound arrives, and our brains can use this effect to estimate the locations of sound sources. That is why people trying to locate a sound tend to turn their heads to and fro while deciding.

Sensing chemicals

An ability to detect specific chemicals in the environment is the most ancient of all the senses, so ancient that even single-celled organisms have a version of it and they use it to detect food, threats, and potential mates. It is a sense that humans seldom use consciously to explore the world. Unlike dogs, for example, most of us do not routinely sniff our way around, although some people with limitations in other senses do. The arrival of a scent is, nevertheless, very commanding of attention and can drive powerful emotional responses, particularly revulsion in response to smells associated with excretion or decay, appetite in response to smells associated with food, and complex emotions triggered by a smell associated with events in childhood. Scent may also play a significant role in sexual attraction, at least at a subconscious level; it certainly plays a major role in sexual and bonding behaviour in many other mammals.

The sense of smell (olfaction) begins in a small region in the very roof of the nose, in specialized cell sheets called olfactory epithelia (Figure 23). Each epithelium is a cell sheet rich in glands that produce a fatty secretion, quite distinct from the high-volume mucus nasal secretions that keep the paper tissue industry in business. The epithelium is also rich in olfactory receptor neurons, located mainly in the epithelium but sending projections above and below it. The downward projections, sticking into the fatty secretion and therefore exposed to any chemicals in the nasal air that can dissolve in this secretion, are called olfactory cilia. The membranes of olfactory cilia contain odorant receptors which, as the name suggests, are able to bind to a range of airborne chemicals that detect as odours.

Humans have a range of different odorant receptor proteins—396 in all—each of which recognizes its own range of broadly similar odorant molecules. Within this recognized set, there will be some

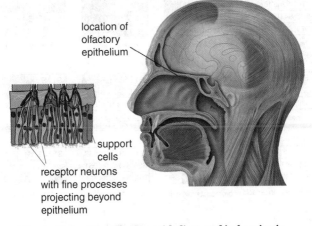

location of
olfactory
epithelium

support
cells

receptor neurons
with fine processes
projecting beyond
epithelium

23. The structure of the olfactory epithelium, and its location in the head.

odorants that bind the odorant receptor strongly and others that bind it moderately or only weakly. Each of the 396 types of odorant receptors has its own set of recognized molecules, but these sets overlap, so that an odorant that is recognized strongly by one odorant receptor may be recognized moderately by two others and weakly by dozens more (Figure 24). When an odorant binds its receptor, the receptor sends signals to activate the olfactory receptor neuron that carries it; the stronger the binding, the stronger the activation. Each one of our approximately 15 million olfactory receptor neurons carries just one type of odorant receptor. This means that each neuron will be activated by the set of odorants appropriate to that receptor, more or less strongly according to how well the odorant binds the receptor.

Above the olfactory epithelium, the olfactory receptor neurons send nerve processes that run to a specific area of the brain called the olfactory bulb. Neurons carrying the same olfactory receptor, even though they may be widely scattered in the olfactory

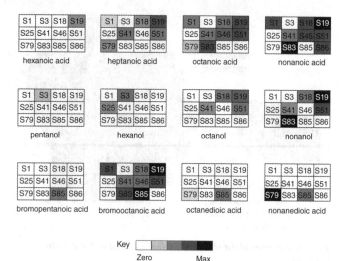

24. **The response of twelve olfactory receptor neurons (with different odorant receptors, S1, S3 ...), to twelve different odorant molecules. The responses of the cells, indicated by darkness of shading, create different patterns in the grid for each molecule. This diagram is based on data which used mouse (not human) tissues.**

epithelium, tend to wire to the same place in the olfactory bulb. This means that the bulb is organized as a kind of 'map', different parts of which will be active in a pattern that is a 'signature' of a particular smell. Figure 24 effectively shows a tiny portion of such a map in its 4 × 3 grids. We still understand very little about how the brain interprets signals on this map in a way that allows us to recognize the perfume of a favourite aunt, to suspect that something is on fire, or to conclude that we must be near to a rotting fish. The most popular assumption is that the brain compares the pattern of activation on this map to memories of patterns that were associated with stimuli, such as aunts or fires, encountered in the past. This might explain how air fresheners, which work mainly by adding new odorants rather than taking unpleasant ones away, seem to remove unpleasant smells; by adding activation areas to the pattern, they are stopping it conforming to a stored memory of something unpleasant.

Our sense of taste is mainly our sense of smell, which is why conditions that compromise smell, such as a temporary head-cold or permanent anosmia due to olfactory nerve damage, seem to render food relatively tasteless or even bizarre-tasting, as the balance of molecules that can reach the olfactory epithelium is altered by congested nasal passages. That said, there are taste receptors in the tongue, of five distinct types that detect five different broad taste types: sweetness, saltiness, sourness, bitterness, and umami (a basic savoury flavour). Some of these receptors are reasonably well understood: for example, the sodium ions of salt cross the cell membrane of saltiness receptors, alter its voltage, and activate the cell. The taste receptors for sweetness and bitterness seem to function in the same way as olfactory receptors in the nose; finding molecules, such as saccharine, that stimulate the sweetness receptors even better than sugar does is the basis of the artificial sweetener industry. In addition to combining olfaction with information from the taste receptors on the tongue, our overall sense of taste also seems to draw on information of texture from the tongue and also the jaw muscles used for chewing. This has been demonstrated in a simple experiment, in which blindfolded volunteers were asked to identify common vegetables served to them either as intact pieces or puréed; their identifications were significantly more accurate when the vegetables were intact.

Light and sight

For humans who can see, sight is the sense with which we most consciously explore and observe the world, to the extent that we use phrases such as 'I see', as metaphors for understanding. Vision begins in the eyes, organs that have a structure similar to a camera in that they focus an image of the external world on to a

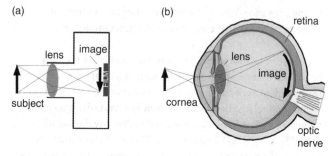

25. Formation of an image (a) in a camera, due to focusing by the lens, and (b) in the eye, due to focusing by the cornea and the lens.

light-sensitive 'screen'. In a camera, the focusing of light is achieved by the lens, a curved piece of glass which, because it has a higher refractive index than air, bends light towards its axis (Figure 25(a)). In the eye, this focusing is performed by two things: the curved front of the eye, which, being full of a high-refractive-index fluid, achieves most of the focusing, and a flexible lens that achieves the rest (Figure 25(b)). The shape of the lens, and therefore its focusing power, can be altered by muscles to allow the eye to focus on objects at infinity (lens relaxed), in the medium distance (lens partially contracted), or close-to (lens fully contracted); the ability of the lens to contract fully is lost with age which is why many older people can be seen reading at arm's length or wearing reading glasses.

The 'screen' of the eye, the retina, is curved (which makes the optics simpler than they need to be for a camera with a flat film). It is a multilayered structure that contains light receptors, neural cells that perform local image-processing, and support cells such as blood vessels. A quirk of the way that the eye develops is that the retina ends up being orientated functionally 'backwards', so that light has to pass through the support cells before it reaches the light receptor cells themselves. One, relatively minor, consequence of this is that the image is slightly more blurred by passage through these support layers than it would be if they were not there. A bigger consequence is that the nerve processes and blood vessels have to traverse the thickness of the retina to connect to the rest of the body; they come together from all over the retina to cross at one point, the optic disc, which in consequence is sightless. We are not aware of this 'hole' in our vision in everyday life, even if we close one eye so that it can no longer compensate. But the existence of the 'hole' is easy to demonstrate. Hold Figure 26 right in front of your face, centred, at a relaxed arm's length. You will be able to see both the spider and the fly. Without moving the paper, look at the fly. Now close your right eye (the one on the same side as the fly), and slowly move the paper towards you, still looking at the fly with your *left* eye. At a particular distance, probably around one foot/30 centimetres, you will no longer be able to see the spider. Interestingly, you will not 'see' the hole in your vision where the spider is hiding (unless you move your gaze away from the fly): your brain will just assume the background tiling pattern continues uninterrupted.

This simple experiment highlights two things: one is that 'seeing' involves a lot of brain activity as well as simple image-collecting in the eye. The other is that our physiology can cause us to miss things 'in plain sight'; the missing spider in the exercise is of no consequence but, if you were about to drive out from a side-road and the 'spider' were a cyclist that your brain just 'painted over' with assumed continuation of background trees, the consequences could be very serious indeed. That is why drivers should deliberately move their gaze across a scene, and not just take one single glance, especially when the vision of one eye might be obstructed by a feature of the car itself, before pulling out of a side-road.

The light-detecting cells of the retina have an unusual structure: a fairly normal-looking cell body and a rod-like outer segment, which is full of in-folded membrane rich in pigment. Different

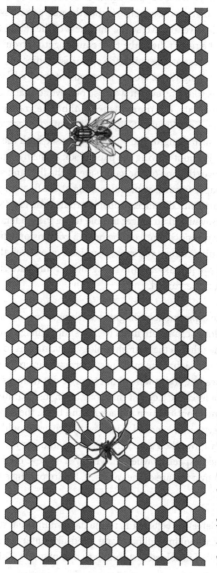

26. A spider and a fly on a tiled floor (instructions for using this image are in the main text).

types of light-detecting cell have subtly different types of pigment, that absorb different wavelengths of light. Cells known as rods contain a pigment called rhodopsin, which can absorb light across the entire spectrum that we normally call 'light'. Cells known as red, green, or blue-detecting cones contain pigments that are sensitive to these colours, and they are the basis of colour vision.

Rhodopsin consists of a protein linked to the chromophore, 11-*cis* retinal ('retinal' is here the name of a chemical, not an adjective). Absorption of a photon of light causes, in about 67 per cent of absorption events, the retinal to flip from its *cis* to its *trans* form, and this in turn causes the protein to change shape (in the other 33 per cent of absorption events, nothing happens—having probable rather than certain consequences is typical of events at the quantum scale). The changed protein activates a chain of events that shuts down ion channels in the cell membrane that normally keep the photoreceptor actively signalling to the next cell along in the optical system, a bipolar cell. Light therefore interrupts this signalling. In this sense, photoreceptor cells may therefore be considered more 'darkness detectors' than light detectors. In our glowing-screen-obsessed society this may seem paradoxical, but detecting darkness may make a great deal of evolutionary sense. The primitive ancestors of eyes may not have been much use for seeing, in the sense that we usually mean the word, but they would still be good enough to signal the arrival of sudden darkness caused, for example, by the shadow of a predator overhead.

The retina is not just a 'camera' for collecting light; it is also a complicated computational device that performs a great deal of image-processing. Within its layers are neural cells and connections that compensate for a whole scene being very bright or dim, and that enhance differences over sameness so that edges within an image are prominent. There are also systems that respond to changes in an image but are quiet when an image remains the same. There are even cells that respond to specific

First, look at the whole figure, and decide on how even the grey colour is in each of the middle sections.

Then, use white paper to cover up the figure above and below a section, and observe the section again.

27. **An optical illusion caused mainly by edge-detection 'circuitry' in the retina.**

directions. The information that leaves the retina to travel along the optic nerve to the brain is therefore not the same in nature as that which may travel along the cable from a TV camera; there is 'picture' information, to be sure, but there is a lot of analytic information added. There is also some information lost, or at least distorted. Each block of grey in Figure 27 is of even intensity, a fact that can be verified if two pieces of paper are used to cover up all but one block. Yet when the whole image is viewed, the blocks look as if they are each a gradient of shading, dark at the top and light at the bottom, because of a 'lie' introduced by the retinal edge-enhancing system.

In all of the systems described in this chapter, detection of a stimulus such as contact, sound, chemicals, and light, is located in the periphery of the body but the full experience of 'sensation' depends on both the pre-processing of information at the sense organ (as in the retinal example above), and subsequent processing of information by the brain. It has long been clear that human perception, unlike the mere data capture of a video or audio recorder, is selective. We do not just see and hear: we look and we listen, and numerous experiments with optical and auditory illusions stress how unconscious processing of sensory data can lead us to draw false conclusions about what we have seen and heard. This has—or ought to have—serious implications for the justice system, in terms of the reliability of witness accounts, and in terms of what can be expected of an accused person such as a pilot, driver, or policeman, who made an unfortunate judgement based on a scene not recorded by an unbiased camera but perceived with the complex, biased, illusion-prone visual system of a human being.

Chapter 5
Reacting and thinking

The central nervous system

The sensory systems described in Chapter 4 send information along nerves, directly or indirectly, to the central nervous system (CNS). In the trunk of the body and the neck, the CNS is called the spinal cord and it is a fairly simple tube of nervous tissue. In the head, the CNS is called the brain and is so thickened and folded in parts that its tubular nature is heavily disguised. Both the spinal cord and the brain can be described in terms of smaller parts, according to anatomy and according to function. In the spinal cord, for example, the most dorsal parts are concerned mainly with receiving sensory information, the middle parts with performing local processing on it, and the more ventral parts are concerned with causing muscles to move (Chapter 6). The brain is divided into many regions, some of which have been clearly identified with functions and some of which are less understood, probably because they cooperate with other areas in performing several different functions.

It is fair to say that, of all the major organs, we understand the brain the least. This is mainly because we are not very good at distilling our most important questions about this particular organ into scientific form. Investigating the physiology of lungs and kidneys is made relatively easy by the fact that we have a clear idea of how to express what they do in scientific terms (the lungs oxygenate blood and remove carbon dioxide from it, for example). It is therefore conceptually straightforward to ask precise scientific questions about how they do this, and to go into the lab and answer them. We can do the same for low-level functions of the CNS, such as the reflex response of withdrawing a hand from something hot, and we therefore understand those functions relatively well (see below). But the most interesting questions, such as 'how do we think?' or 'how am I able to read this sentence?', are very hard to translate into scientific terms: how does one even measure the amount of thinking that is happening? The slow progress made in understanding these complex functions of the CNS sometimes encourages a kind of mysticism about the brain amongst non-physiologists, and a belief that something very 'different' is going on at the level of its cells. There is no physiological reason to believe this to be the case; when we examine the cells, there is nothing about any of them that does not fit normal physiological understanding. Our main problem in understanding 'thinking' is that we are have not yet worked out how to express our questions in a way that scientific experiments can clearly answer.

Cells of the CNS

The CNS is dominated by two cell types; neurons and glia. It also has the normal support systems, such as blood vessels, present in most tissues, but it is the glial cells and neurons that play the most direct roles in processing signals. Glial cells were originally thought of merely as support cells for neurons (the word shares a root with 'glue'), and they certainly do provide many types of mechanical, biochemical, electrical, and defensive support. It has recently been shown that some types of glia can play an active role in signalling within the CNS, but most signalling is by neurons.

Anatomically, neurons are unusual cells that contain long processes. A typical neuron bears a tree of fine processes called

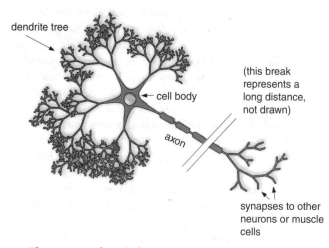

dendrite tree

(this break represents a long distance, not drawn)

← cell body

axon

synapses to other neurons or muscle cells

28. The structure of a typical neuron.

dendrites, which act as receivers of information, and a long, thick axon along which information processed by the neuron passes toward other cells (Figure 28). Many axons are only a few millimetres long but some are long enough to stretch, for example, from the small of your back to the furthest muscle of your foot. When outside the CNS, and sometimes inside it, axons tend to exist in bundles. Inside the CNS, these bundles are called tracts, and outside it they are called nerves.

Neurons have evolved to be efficient in transforming chemical information into electrical signals, and then processing these signals and passing them on. Electricity in cells is the same as any other electricity, in the sense that it relies on the unequal distribution of charge (voltage), and the movement of charge (current). In ordinary copper wires, the charge is carried by electrons and it moves along the wires in the same direction as information coded by the electrical current. Cells, however, arrange their electrical currents very differently. Charge is carried by positive and negative ions dissolved in cellular fluids, and the

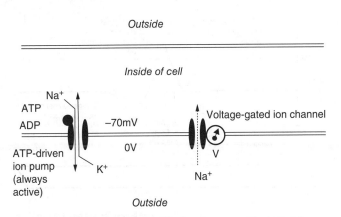

Outside

Inside of cell

Na⁺

ATP

ADP   −70mV   Voltage-gated ion channel

0V

ATP-driven   V
ion pump   K⁺
(always   Na⁺
active)

Outside

29. Ion flows across the cell membranes of neurons. The ATP-driven pump continuously expels sodium from the cell and imports potassium, making the inside of the cell −70 mV compared to the outside. This voltage can be shorted out by voltage-gated ion channels.

direction of current flow is across the cell membrane at right angles to the direction of information flow along axons and dendrites (Figure 29).

In a resting neuron, the activity of ATP-driven ion pumps in the cell membrane generates a voltage such that the inside of the membrane is at a voltage of about −70 mV compared to the outside. The membrane also contains voltage-dependent channels. These open if the membrane voltage rises to about −55 mV and, when they open, the channels allow ions to flood back across the membrane and make its voltage rise even more, all the way to positive. A short time later, the channels automatically close and enter a temporary hiatus during which they cannot be opened again. This allows the constantly working ATP-driven pumps to restore the voltage to its resting state. Therefore, if something causes a local area of membrane to become less negative than about −55 mV, it will experience a spike of positive-going voltage, before returning to its resting state of −70 mV (Figure 30).

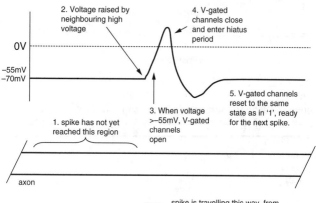

30. A spike travelling along an axon from right to left, with the voltage at different points plotted on the graph above (the graph can also be read as a sequence of events at one point, with time running left to right). The small dip of voltage below −70 mV is an effect of potassium ion flows, not described in the main text.

Furthermore, a spike in one part of the membrane will cause the voltage to rise above −55 mV in the immediately neighbouring parts of the membrane, so that part will spike too. This will cause the next part of the membrane to spike, and so on. The net effect is transmission of the spike along the membrane, its height never diminishing because it is always powered locally from the ion pumps.

The above mechanism allows neurons to transmit electrical pulses along their axons, but how do the pulses start in the first place? In the dendrites of neurons, there are receptor molecules that respond to neurotransmitters, messengers secreted by (mainly) other neurons. Some receptors respond to binding a particular neurotransmitter by opening membrane channels and allowing the local voltage to rise. Some receptors, for different neurotransmitters, have the opposite effect or may trigger cascades of chemical signals in the neuron. The various voltages originating in different parts of the dendrite, and the chemical signals, combine together in a region of the cell called the axon hillock, which is effectively the boundary between the 'receiving' end of the cell and the 'transmitting' end. The axon hillock has a high density of voltage-sensitive channels and, if the voltage of its membrane is sufficiently high, it will initiate spikes that will travel along the axon in the manner described in the last paragraph. Typically, occasional spikes will happen anyway, but they become much more frequent when the balance of incoming signals is positive. The neuron therefore encodes the subtle balance of all the inputs it receives as the frequency with which spikes travel along its axon.

At the end of the axon is a synapse: a specialized area which is full of stored neurotransmitter. The synapse is located a very short distance from either the dendrite of another neuron, a muscle, or a hormone-producing cell (see Chapter 6). Spikes coming along the axon cause the release of tiny amounts of neurotransmitter; the more frequent the spiking, the more neurotransmitter is released. The dendrite of the receiving cell will have receptors that respond to the neurotransmitter, and this will be one piece of information that the receiving cell will integrate with information coming from many other synapses, to determine how frequently it will generate its own spikes, and so on. Each axon can branch to lead to several synapses. The neurons of the CNS therefore form a vast network in which information is split, combined, and somehow processed.

We understand some cases of the processing reasonably well; examples include reflex arcs (described below), the 'circuitry' that detects features such as edges in images coming from the eyes, and simple types of memory. Most other things in the brain, especially those we are most aware of subjectively such as thoughts and feelings, are not yet understood at all well. Certain drugs, both medical and 'recreational', work by inhibiting or stimulating specific types of synaptic signalling and have well-known effects on mood and feelings of reward. This observation often results in sloppy thinking, rife in 'pop neuroscience', that equates the pathway affected by the drug with the normal version of the mood, and creates the illusion that we now understand the neural basis of the mood. This conclusion is no more sound than would be the conclusion that, because one can no longer hear a symphony on one's headphones when their lead has been cut, the lead must be the physical explanation for the symphony.

The reflex arc

If you make the mistake of touching something unexpectedly hot, your arm will pull back automatically. This withdrawal is literally faster than thought, because it uses a very simple 'loop', a reflex arc, that involves very few neuron-to-neuron connections and is therefore relatively easy to analyse and understand. A famous example of a reflex arc is the 'knee jerk' that can be elicited by sudden stretching of the patellar tendon. Doctors often test the reflex by tapping the tendon in a sitting patient, as a simple way to test some basic neural functions. Here a mechanical stretch receptor in the quadriceps muscle that is attached to the tendon stimulates a sensory neuron whose cell body lies in a dorsal root ganglion connected to the spinal cord (Figure 31). This neuron sends an axon to make an activating-type synapse with a motor neuron in the ventral part of the spinal cord, and this motor neuron sends its axon back to the quadriceps muscle, causing it to contract. This contraction extends the knee joint, causing the classic 'knee jerk' in an individual with a freely dangling leg. At the same time, a slightly longer path (Figure 31) sends signals that inhibit contraction of the hamstring muscle, stopping the two muscles fighting one another. In ordinary life, the reflex does not produce the wild jerks seen in response to an artificial stimulus to an unloaded leg in a doctor's surgery, but is part of a system that automatically corrects for any tendency for our legs to buckle under us as we stand and move.

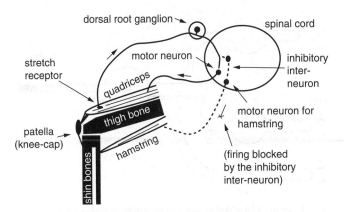

31. The 'knee-jerk' reflex arc.

Elementary learning and memory

One of the most elementary examples of learning is the development of an association between two different sensations, for example the sound of a spitting frying pan and the pain of being burned by droplets of hot fat landing on the hands. Experienced cooks tend to pull back in a reflex-like way when they hear the sound. In the late 1940s, the Canadian physiologist Donald Hebb suggested a mechanism that could account for this: he proposed that the strength of a synapse between a sending neuron and a receiving neuron would increase if the receiving neuron fired just after the sending neuron signalled to it. Consider an idealized system (Figure 32), in which neuron S, stimulated by the sound of spitting fat, and neuron P, stimulated by pain, each synapse with neuron A, which triggers some kind of aversion response such as withdrawing hands from the region of the pan. In an incautious beginner cook, the firing of neuron S will often be accompanied by firing of neuron P, so A is often fired by P's pain signal a fraction after S fires. The synapse between S and A therefore becomes stronger and stronger, until eventually the firing of S alone is enough to cause A to fire. The effect is called long-term

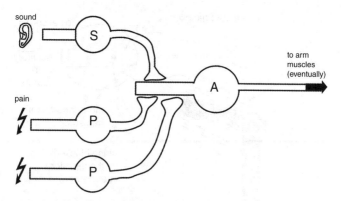

32. The layout of the 'frying pan' example used in explanation of long-term potentiation in the main text. This is a simple example to illustrate an idea; it does not represent real neural anatomy.

potentiation; effectively, the cook has learned to associate the sound with the pain and now shows the hand withdrawal response as soon as the sound is heard without waiting to feel pain.

In the years since Hebb made his proposal, work on animal systems has identified real examples of Hebbian learning, particularly for synapses that use the neurotransmitter, glutamate. Post-synaptic ('receiver') cells have two types of glutamate receptor. One type, AMPAR, act in the normal way and trigger a path that acts to stimulate the neuron (Figure 33). The other type of receptor, NMDAR, is voltage-sensitive. If the receiving cell membrane's voltage is still near its inactive state, they show little activity even in the presence of glutamate. If, however, the receiving cell membrane's voltage is a lot less negative, as it would be if the cell were being stimulated by inputs from other neurons as well, the NMDAR receptors are active and trigger biochemical changes that increase the efficiency of signal transmission from the AMPAR receptors. Thus neurons that receive inputs of several

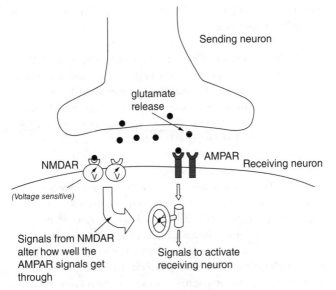

Sending neuron

glutamate release

NMDAR

AMPAR Receiving neuron

(Voltage sensitive)

Signals from NMDAR alter how well the AMPAR signals get through

Signals to activate receiving neuron

33. Molecular mechanisms of an example of long-term potentiation.

different types will, if the two stimuli keep occurring together, increase their sensitivity so that just one stimulus might be enough to trigger the response.

A concrete example in a simple animal is provided by fruit flies. If flies are given an unpleasant electric shock in the presence of an otherwise unalarming odour, they will activate escape behaviours. Later, the presence of the odour molecule alone, with no shock, is enough to trigger the escape behaviour. But, if the experiment is done with flies in which the genes that encode the NMDAR receptors are inactivated, those flies fail to learn the association between odour and shock. Studying this type of learning in humans obviously runs into technical and ethical difficulties (the frying example was just a made-up illustration of the idea), but it has been possible to identify long-term potentiation, of the same type, in small pieces of living brain removed from patients as part of brain operations.

Clearly, there is a great difference between learning an association between a sound or smell and pain, to create a conditioned reflex, and learning in the sense of learning the seven times table, or learning to play the viola, or learning how to design experiments to discover things not previously known. A significant number of neurophysiologists assume that long-term potentiation is the basis even of 'higher' learning events, and that the higher-level information (facts, faces, behaviours) is represented by vast networks of synapses. The word for the thing encoded by such a network is an 'engram'. To be clear, nobody has yet 'seen' an engram, but a combination of direct animal experiments and clinical observations on humans undergoing brain surgery has suggested that stimulation of neurons in diverse areas of the brain can reproducibly activate memories. Similarly, inhibition of firing in specific areas can temporarily remove a memory (at least for very simple animal examples, mainly conditioned reflexes). One of the slightly surprising things about the brain is that pieces of it can be removed, for example in surgery for brain cancer or epilepsy, and, while the patients may seem a little slower in recalling things generally, they seldom lose specific memories. Again, this suggests that most memories do not exist in one single place but are distributed in large networks that are tolerant of loss. All of these observations are compatible with the engram idea, and emphasize how much brains differ from the electronic computers with which they are too often equated.

Some higher functions, such as spoken language, can be associated strongly with specific regions of the brain. Historically, correlations were made by studying the brain functions lost in patients who have suffered anatomical losses of brain tissue due to local strokes or traumatic injury. More recently, the brain-scanning technique called functional magnetic resonance imaging (fMRI: Chapter 1) has enabled researchers to study which parts of the brain are most active when someone is performing specific mental tasks, such as recognizing faces or solving spatial problems. Examples of associations between functions and brain areas include control of heartbeat and breathing (brainstem), regulation of thirst, appetite, temperature, sleep, and emotional state (hypothalamus), coordination of voluntary muscle movement (cerebellum), perception, initiation of movement, language, and personality (cortex).

Some important systems, for example the limbic system that controls emotion, arousal, and reward, include many brain parts. Disorders of the limbic system are associated with a range of psychiatric conditions (e.g. depression, bipolar disease) and the system is therefore a common target for drugs designed to control these diseases. The drugs generally work by interacting with the molecular machinery of synaptic transmission to increase or decrease the effective activity of particular neurotransmitters: depression is often treated by increasing the strength of serotonin-mediated signals, for example. Because of its association with rewards, the limbic system is also often the target of 'recreational' drugs, many of which (e.g. heroin, cocaine) strongly activate parts of the system that use dopamine as a neurotransmitter, sometimes ten times more strongly than they are activated by natural euphoria-inducing triggers, such as sex. Unfortunately, over-stimulation makes the system less sensitive, meaning that natural behaviours are no longer rewarding and people become dependent on the drug, needing higher and higher concentrations to achieve what were once everyday feelings of well-being.

The 'highest' of all brain functions is consciousness, something we recognize in ourselves and others but that is not easy to define clearly. From the fact that damage to the brain causes damage/alteration to the conscious personality, and the fact that there seems to be no material alternative, the consensus assumption of most physiologists/neuroscientists is that consciousness (the mind's 'I') must be an emergent property of neural activity in the brain. At the moment, this remains an unproven assumption. For the brain, at least, human physiology still feels like a young science with a great deal of developing still to do.

Chapter 6
From thought to action

Whatever the mechanisms of the mind's internal cogitations, the ultimate result of sensation and thought is usually some kind of action, be it moving the whole body, manipulating an object with the hand, moving diaphragm, mouth, tongue, and voice-box to speak, moving fingers to type a sentence about physiology, or perhaps merely blushing. All of these depend on muscles which, in their various forms, provide a nearly universal means for the nervous system to control the body and the world.

Muscle: a tissue specialized to pull

Muscle cells are highly adapted for turning chemical energy into mechanical force. They use molecular machines that are present in almost all cells for generating small, local forces and movements, but muscle cells arrange these components in a special way that enables each tiny contribution to be added together. This is true of all types of muscle cell but is seen most clearly in skeletal muscle—the type of muscle responsible for most voluntary motion (and some involuntary motions, such as reflexes, and breathing).

The cells of skeletal muscle are unusual in many ways. To begin with, they are huge, being formed by the fusion of many immature cells to create long, cylindrical tubes that contain many nuclei.

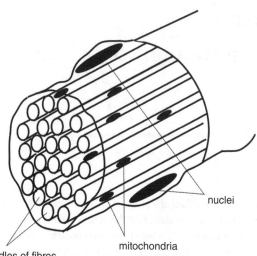

nuclei

mitochondria

bundles of fibres

34. The internal structure of a skeletal muscle cell: the bundles of fibres contain the stacked protein filaments shown in Figure 35.

Then, these nuclei are not in the middle of the cell, as they are for most cell types, but are forced to the outside edges to make way for a vast, parallel array of protein bundles that dominate the cell's inner core (Figure 34). The protein bundles consist of filaments of two intimately mixed types, thin and thick. The thin filaments consist of long polymers of a relatively inert protein called actin, with a few accessory proteins bound along the filament's length. Actin filaments project outwards in both directions from a so-called Z-line (also made of proteins). There are many Z-lines in each cell so that, as far as actin is concerned, the sequence is actin filaments, Z-line, actin filaments (pointing the other way), then a gap, then the pattern repeats (Figure 35(a)). The thick filaments have a broadly similar structure, in that they consist of filaments projecting one way on one side of a central line (in this case, the M-line), and pointing the other way the other side of it (Figure 35(b)). They consist of entwined molecules of a protein

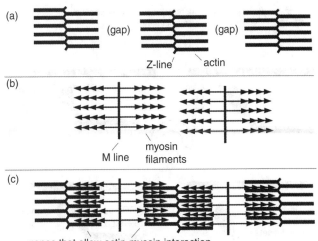

(a)

(gap)

(gap)

Z-line

actin

(b)

M line

myosin filaments

(c)

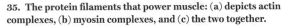

zones that allow actin-myosin interaction

35. The protein filaments that power muscle: (a) depicts actin complexes, (b) myosin complexes, and (c) the two together.

called myosin. The Z-lines and their associated thin filaments alternate with the M-lines and their thick filaments, so that the actin and myosin filaments interdigitate (Figure 35(c)). This interdigitation is critical for muscle action.

As might be supposed from their arrangement, myosin and actin interact, and they do so in an interesting way. Myosin proteins have heads that are located just beyond a movable hinge region of the protein, and these heads can bind to ATP, the energy-carrying molecule described in Chapter 2. Myosin heads attached to ATP can also bind to actin and, when they do so, their ATP is hydrolysed to ADP and phosphate. As long as there are calcium ions available to permit helpful interactions with accessory proteins, the ADP and phosphate are released by the myosin heads and this release causes the hinge of the protein to bend. The bending moves the head, and therefore pulls on the attached actin

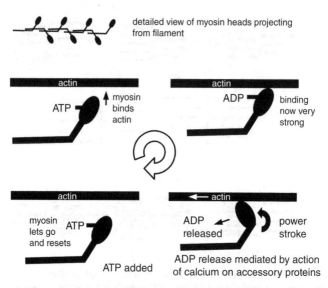

detailed view of myosin heads projecting from filament

actin

ATP ↑ myosin binds actin

actin

ADP binding now very strong

actin

myosin lets go and resets ATP

ATP added

actin →

ADP released power stroke

ADP release mediated by action of calcium on accessory proteins

36. The cycle of myosin binding to, pulling on, and releasing actin. The asynchronous pulling by many myosin heads stops the actin slipping back when some heads are in the release phase.

with a definite force, albeit a small one (around a million-millionth of a Newton). The myosin remains tightly bound to actin until another molecule of ATP encounters it, when the head will bind the ATP and let go of the actin, and the hinge will extend again ready for the cycle to repeat (Figure 36).

Along an active muscle, this cycle is repeated over and over again across the trillions of myosin heads. They do not all bind and contract at exactly the same time, some myosins holding on while others let go and extend, ready to pull again soon. Thus, on average, the system acts as a ratchet and pulls the actin filaments along against the myosin filaments. If the force being developed exceeds the resistance to be overcome, the muscle physically contracts, moving the attached part of the body. If the force just balances resistance, no actual muscle shortening takes place but the force can still be applied to balance an external force as when, for example, one holds an apple against gravity in the palm of an outstretched hand. An apple bears down with a force of one Newton, so a quick calculation indicates that the force of a million million myosin heads must be being brought to bear at any one time just to hold it in that position. Even this is an underestimate, because the lever action of our limbs (see later) means that arm muscles have to pull about ten times harder than apples bear down on the hand.

The fact that myosins grip actin hard, and require ATP to let go, has an interesting consequence well known to pathologists and readers of crime novels. When someone dies, their muscle cells will no longer be supplied with fresh oxygen and food and their ability to generate new ATP will be lost. The myosin heads of all of their muscles will therefore be unable to let go of actin, and the muscles will therefore become completely unyielding. This state, rigor mortis (Latin for 'stiffness of death'), begins about four hours after death at typical room temperatures, peaks at around twelve hours, and persists for around two days, after which decomposition, mainly by cellular enzymes, breaks the integrity of the filaments and the remains of the muscles can relax. This long ongoing biochemistry, one example of many, challenges the concept that death consists of a single moment at which the body goes from living to dead.

In the account of myosin activity above, I mentioned the need for calcium ions to be present for accessory proteins to assist the action of the myosin head. This requirement is so absolute that the body uses it as the mechanism to control the activation of muscles and to determine how hard they work. Muscle cells have an extensive internal network of membrane-bound cavities called the sarcoplasmic reticulum, which ramifies around the actin and myosin bundles. The sarcoplasmic reticulum is essentially a calcium store; it is equipped with systems to scavenge calcium ions from the cell and sequester them, and its membrane is equipped with channels that can open and allow the calcium to flood out and reach the actin-myosin fibre systems to activate them.

The outer membrane of the cell is rather like that of a neuron, in that it maintains a resting negative voltage but it can propagate a voltage spike if one is induced in it. Axons from motor neurons, such as the motor neurons described in the passage on the knee-jerk reflex in Chapter 5, make synapses with muscle cells. Spikes coming along those axons induce, via the release of neurotransmitters at the synapse and their interaction with receptors on the muscle cell, voltage spikes in the muscle cell membrane. Parts of the muscle cell membrane fold inwards and bring the voltage spikes inside the cell with them. There, protein complexes mediate communication between the infolded membranes and the sarcoplasmic reticulum. Voltage spikes activate the protein complexes and cause the sarcoplasmic reticulum to release some of its stored calcium, which then activates the myosin (Figure 37). The more frequently the spikes arrive, the greater the fraction of time the calcium channels are open, the greater the concentration of calcium at the filaments, and the harder the muscle cell pulls. Muscles consist of large numbers of individual muscle cells, each of which will be stimulated by synapses from motor neurons, and the overall force developed by the muscle will be determined by an average of the activities across all its constituent cells.

Some drugs and toxins work by interfering either with muscle action or with the ability of motor neurons to signal to muscle cells. A famous example is Botulinum toxin, produced by the bacterium that causes the disease botulism. This toxin is taken up by the ends of motor axons and blocks their ability to release neurotransmitters, causing paralysis. It has some medical uses for blocking inappropriately active muscles in some diseases, and is also used cosmetically as a relaxant of facial muscles, reducing wrinkles, under trade names such as 'Botox'. Curare, a plant-derived

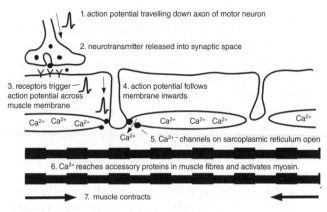

1. action potential travelling down axon of motor neuron

2. neurotransmitter released into synaptic space

3. receptors trigger action potential across muscle membrane

4. action potential follows membrane inwards

Ca^{2+} Ca^{2+} Ca^{2+} Ca^{2+} Ca^{2+} Ca^{2+} Ca^{2+}

Ca^{2+} 5. Ca^{2+-} channels on sarcoplasmic reticulum open

6. Ca^{2+} reaches accessory proteins in muscle fibres and activates myosin.

7. muscle contracts

37. The sequence of events that couple the firing of motor neurons to activation of muscle contraction.

poison, blocks the ability of the receptors on the muscle cells to detect released neurotransmitters. It therefore induces paralysis in prey hunted with poisoned arrows (the meat can be eaten safely afterwards because curare is not absorbed from food), and has seen use, in carefully controlled doses, as a muscle relaxant in surgical anaesthesia.

The musculoskeletal system

The internal structure of muscles has evolved to enable them to pull, yet a great many—probably the majority—of our everyday actions involve pushing. We walk or leap forward by pushing against the ground, we push food on to our forks and into our mouths, we push plugs into sockets, and roughly half the doors we open in our lives, we push. The body clearly needs a mechanism to transform a pulling force into a pushing one and, in fact, it has two; one is based on mechanics, and the other is based on hydraulics.

A pivoted lever is a very simple mechanical device that can be used to transform pulls into pushes and vice versa. The oar of a rowing boat is a simple example: when using oars in the conventional way, the rower pulls on the handle of the oar, which pivots at the rowlock, and the blade at its far end pushes on the water. If the pivot is halfway between applied force and load, then the force of the push will be equal to that of the pull. If the pivot is closer to the source of pull, as in oars, the push will be less hard than the pull but the pushing end of the lever will move more (so that force × distance moved is the same at both ends). It turns out that muscles can provide much more force than we need to exert on our environment (e.g. to lift a cup of coffee), but we typically need to move things over a much greater distance than a muscle can contract (moving that cup all the way from table to mouth). For this reason, muscles are usually attached close to the pivot point of the levers they operate, while the load is applied much further away.

The levers are, as one would expect, almost rigid structures. In humans, they are made of mineral-rich, but still living tissue: bone. Muscles attach to bones via specialized adaptor tissues, tendons, which do not contract actively as muscles do but are somewhat flexible and make very strong connections with both tissues. Bones are also connected to one another, around joints, by fibrous structures very similar to tendons, called ligaments. Tendons and ligaments are capable of carrying extremely high tensions and, before advanced metallurgy produced reliable springs, animal tendons were wound into sinew rope and used in machines that involved high tensions, such as the drive springs of rock-hurling siege engines. Together, muscles, tendons, ligaments, bones, and some special joint structures between the bones form the musculoskeletal system, which is responsible for almost all of our externally observable movements (the exceptions being expulsion of fluids and babies, and penile erection, all considered later in this chapter).

The arm provides a convenient illustration of the way in which muscle and bone work together, and it is accessible to self-experimentation. The most obvious joint in the arm is the elbow, a simple hinge between the upper arm and the forearm, which leads to the hand. The main muscles that operate the hinge are the biceps and the triceps (the names mean 'two-headed' and 'three-headed' respectively, and refer to their anatomical shape). If you stand or sit with your elbow by your side and right forearm outstretched, palm upward, then the biceps will lie at the front of your arm and the triceps at the back (Figure 38). The biceps runs down from the shoulder area to connect to the radius, the main bone of the forearm beyond the elbow joint. Contraction of the biceps therefore acts to flex the elbow and raise that outstretched hand. You can feel its action by gently gripping, with your left hand, the middle of your upper right arm. If you place your outstretched right hand under something heavy, for example a table, and try to push the table up from the floor, your left hand will feel the biceps muscle in the front of your right arm thicken in girth as it contracts as much as it is able to.

The triceps muscle runs down the back of the upper arm (assuming the arm is still in its elbow-by-side, hand-out, palm-upward position) and inserts on the very back of the ulna (the other forearm bone), from around the back of the hinge-point of the elbow (Figure 38). Contraction of the triceps therefore acts to extend the elbow and straighten the arm. Again, you can feel its action by having the same grip as above, but this time having the

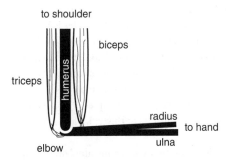

38. Muscles that flex and extend the elbow.

hand on top of the table and trying to push the back of your hand down through the table. Your palm still needs to be upward for that to work (if it is not, both the biceps and triceps take a somewhat spiral route and, while you would still be able to feel them contracting in these circumstances, it would be harder to make sense of what you feel). The fact that the biceps and triceps attach (via tendons) to bones of the forearm so close to the elbow emphasizes how lever action is used to magnify movement at the hand end, at the expense of diminished force compared to what is developed directly by the muscle.

Because the action of the biceps is to flex the elbow and the action of the triceps is to extend it, the muscles can be considered to work in opposition to one another. This is a general principle of the musculoskeletal system: one muscle or group of muscles generates one movement and another muscle or group of muscles generates the exact reverse movement. In the knee-jerk reflex described in Chapter 5, the main reflex pathway that caused contraction of an extensor muscle was accompanied by a pathway that inhibited contraction of the corresponding flexor muscle precisely because the movement and position of the shin results from a balance of the two activities, so the flexor had to be inhibited for a robust extension to occur. In a healthy and awake person, limbs do not just dangle but are held in a specific place by the actions of opposing muscles, working just hard enough to maintain the position of the limb against environmental perturbations. This gentle continuous action of muscles is known as muscle tone. If we expect to have to maintain a limb's position against a large force, or to move with high precision, we can choose to have a very high muscle tone (ballroom dancers consciously do this with their arms and shoulders, for example, and call the resultant robust shape their 'frame'). The principles of opposition and tone are equally true of the muscles of the trunk and head.

For many muscles, whether they are used for 'pushing' or 'pulling' depends on context rather than on their own mechanics. The biceps, for example, might be the main muscle used when flexing the arm when 'pulling' an apple off a tree, or when flexing the arm to 'push' a vertically sliding sash window upwards to open it. For this reason, physiologists tend to classify muscle action by reference to flexion or extension of a joint, a purely internal thing that does not depend on the external use to which the movement is put.

The biceps and triceps are not the only muscles in the upper arm; they work together with other, smaller muscles in a way that allows a choice, for example, of pushing down on the table with the back of the hand, its side, or the palm. Our vast range of available movements, strong enough to hurl a javelin and precise enough to thread a needle, emerges from the relatively large number of muscles we have for each main joint, and from very fine and separate control not only of different muscles but often of different parts of those muscles, by the central nervous system.

Squeezing cavities

Not all muscles are associated with the skeleton and with pulling on bones. In many parts of the body, muscles are arranged circumferentially around a cavity, and their contraction compresses the contents in that cavity, causing them to move to somewhere that is free to expand to accommodate them or to move out of the body altogether. Two examples of this described in Chapter 2 were the heart, which pumps blood, and the gut which pumps food or, towards its end, faeces.

The absorptive epithelium of the gut tube is surrounded on its outside by two layers of muscle. These are of a type called smooth muscle, different in a number of details from skeletal muscle but the basic mechanism for turning chemical energy into force is the same. In the inner layer, the muscle tissue is arranged to run around the circumference of the gut tube, as barrel hoops run around a barrel; these are therefore called circular muscles. In the outer layer, the muscle tissue runs along the tube and is called longitudinal muscle (Figure 39(a)). When the circular muscle contracts at a particular point along the gut tube, it reduces the diameter of the gut at that point and increases its length. When the longitudinal muscles contract, on the other hand, the gut tube becomes shorter but wider. Again, there is a principle of opposition, two muscle types having opposite local effects, but in this case the opposing muscles collaborate by acting in sequence. A slow wave of contraction of longitudinal muscles and relaxation of the circular muscles travels along the gut, in the stomach-to-anus direction, immediately followed by a relaxation of the longitudinal muscles and contraction of the circular muscles. Some waves travel a long way and some only a short way, but their local effect is that a construction of the lumen travels along the gut, pushing food forward into an open area beyond (Figure 39(b)).

Presence of food within the gut is itself a stimulus for the initiation of a wave, so that food itself triggers the systems that move it. Coordination of the muscles is somewhat complicated, relying both on local electrical communication between muscles and on the action of an extensive nervous system that is located mainly along the gut itself, with some inputs from the central nervous system. This gut, or 'enteric' nervous system, is an excellent example of a very large aspect of complicated neural activity over which 'we', as conscious beings, have almost no control (we do have some control over events at the two ends of the gut). Indeed, except for the odd rumbling noise or churning sensation, we do not even have any knowledge of what the enteric nervous system is doing. This is by no means unusual: bodies are full of neural systems that run their physiology without reporting in any subjectively discernible way to the conscious mind, to the extent that knowing that these neural systems exist at all had to await the age of science.

Blood is also pumped by the action of muscles, mainly those of the heart (though muscles in the large vessels also play a role in

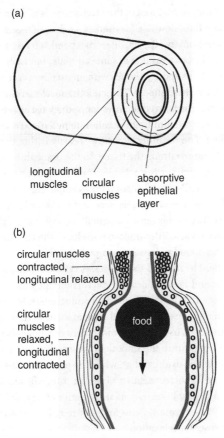

(a)

longitudinal
muscles   circular
muscles

absorptive
epithelial
layer

(b)

circular muscles
contracted, ———
longitudinal relaxed

circular
muscles
relaxed, ———
longitudinal
contracted

food

39. Structure and action of gut muscles: (a) shows the inner layer of circular muscles and outer layer of longitudinal muscles; (b) shows longitudinal muscles contracting ahead of food to open the gut to create space, while circular muscles contract behind the food to push it forwards.

controlling the pressure and speed of blood flow). In most of these smooth and cardiac muscle systems, movement of fluid is the ultimate goal and movement of muscles is used as a means to achieve it. Sometimes, though, the mechanism is used in the other direction, and fluid flow itself is used as a driver of tissue movement. In humans, the hardness of a penile erection is mediated not by bone (as it is in some animals) but hydraulically. The penis contains spongy tissues with an excellent connection to a high-pressure blood supply. The smooth muscles around the small arteries (arterioles) that enter the spongy tissue are normally contracted, limiting the rate at which blood can flow through them. The resulting slow flow is easily within the capacity of the small veins draining the tissue, so the tissue does not experience any high pressure. Under conditions of sexual excitement (or REM sleep, or sometimes apparently randomly, especially in adolescents), nerves serving the arterioles are activated and their ends release a signalling molecule, nitric oxide. The nitric oxide causes the smooth muscles of the arterioles to relax, allowing blood to flow in unimpeded and to inflate the spongy tissue. The inflated tissue presses on the drainage veins, closing them and raising pressure still further, making the previously soft and flexible tissue rigid and straight. In this way, fluid motion and pressure, derived ultimately from the contraction of heart muscles, is used to drive movement of a bone-free organ. Enzyme systems within the smooth muscle cells slowly deactivate the effects of nitric oxide so that, when the nervous input reduces, the muscles will contract again and the erection will subside. The drug Sildenafil, best known by its trade name Viagra, works by inhibiting these enzyme systems and so strengthening and prolonging vascular relaxation.

Should a reader have been embarrassed by the last-mentioned topic, they may have indicated it by blushing. The reddening of facial skin that constitutes the blush is caused by increased blood flow through capillaries near the surface of the skin, and that in turn seems to be caused, again, by nervous modulation of blood vessel muscle tone. Both physiological systems, erection and blushing, are subjectively strange because we have no direct control over them in the way that we have control over the movement of our fingers. We have an awareness, through experience, of conditions likely to activate the mechanisms but can neither activate them nor prevent activation by conscious control. Mechanisms like this remind us that, while the mind may be an emergent property of the nervous system, that system does a great deal that is beyond the reach of the mind. Some unconscious actions, like heartbeat and food peristalsis, are mediated directly by nerves going to effector tissues. Other unconscious actions use nerve transmission only part of the way, neurons passing signals on to cells that produce hormones that spread throughout the body in the blood. We have met some hormones, such as insulin, earlier in this book and will meet others, such as sex hormones, later. They are mentioned here just to stress that, while muscle contraction is the most obvious way in which the nervous system can cause something to happen, it is not the only mechanism available.

Chapter 7
Defence

The inside of the body is warm, wet, nutrient-rich, and has all parameters highly controlled. Outside is a highly variable environment, with strong fluctuations in temperature, humidity, wind, and sun. The body therefore has to invest considerable effort in maintaining a barrier to protect its internal conditions from heat, cold, and desiccation. It also has to protect itself from being rotted alive by micro-organisms that would enter and colonize it, given the opportunity.

The integumentary system

The integumentary system consists of the skin and its various appendages such as hair and nails. The skin is the largest organ of the body and is composed of three main layers (Figure 40). The innermost layer, the hypodermis, consists mainly of fat cells that serve both as stores of food and as thermal insulation. Outside that is the dermis, a thick layer of connective tissue that provides strength and elasticity; this layer is well served by blood vessels and sensory nerves. The outermost layer, the epidermis, is itself composed of sub-layers. In the layer nearest the dermis is a population of stem cells, which divide slowly and give rise both to replacements for themselves and to progeny that enter the next layer up. In this layer, cells divide repeatedly and then become more specialized as they are pushed outwards by more ongoing

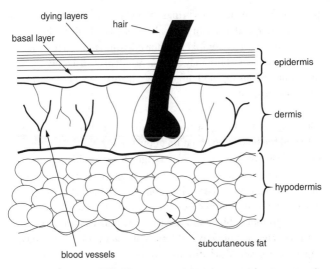

dying layers

hair

basal layer

epidermis

dermis

hypodermis

blood vessels

subcutaneous fat

40. The structure of the skin.

division in the layers below. During their specialization, they fill with a fibrous, water-retaining protein, keratin, stick together firmly, and lose their nuclei. In this state they form the visible, outer layer of the skin. Eventually, the cells flake off as dust, and are replaced by new ones coming up from underneath.

The outer layer of the skin is highly impermeable, both to water (in either direction), physical objects, and micro-organisms. It does harbour an extensive ecology of bacteria on its surface, but these live mostly in a symbiotic relationship, fed by skin secretions and dead cells. They cause no problems on undamaged skin, and help to defend the body by secreting substances that kill other micro-organisms, or at least prevent them from thriving.

Damage repair

The skin is strong but it can still be breached, for example by a sharp thorn, which may well penetrate to at least the dermis and possibly even further. This brings immediate problems of a failure of the barrier, with consequent leakage of body fluids to the outside and with a potential path for micro-organisms to invade from the surface; there may also be problems caused by micro-organisms injected directly by a dirty thorn, but we will consider those later.

The body's immediate priority, when the integument is breached, is to seal the hole by coagulation of the leaking blood. Healthy blood always has the potential to coagulate but, as long as it is flowing in undamaged vessels, it is not triggered to do so. If the lining of vessels is damaged, however, components of blood come into contact with cells and other material of the tissue beyond. When a protein in blood called Factor VIIa comes into contact with cells that surround blood vessels, it is activated, and in turn triggers a cascade of activations of other blood factors. After several steps, these turn a soluble blood protein, fibrinogen, into the insoluble fibrous protein, fibrin. Other pathways can also turn fibrinogen into fibrin, but the one described above seems to be the most important in everyday life. As this sequence of events involving soluble proteins is going on, a separate mechanism recruits defensive cellular materials. Blood always contains small cell fragments, platelets, created by the break-up of a specific type of parent cell in the bone marrow, this break-up being part of its natural development. Platelets ignore the lining of a healthy blood vessel, but if this lining is damaged, they stick to the exposed underlying structures and are activated by this adhesion. The activation increases the platelets' stickiness further and transforms them from being smooth, rounded cells to star-shaped ones that will easily tangle with one another. It also causes them to release a factor that activates other nearby platelets, and promotes inflammation (see later). The combination of fibrin fibres tangling to make a gel, and star-shaped activated platelets tangling with each other and with the gel, generates a clot strong enough to stem blood flow (from a minor wound) and to provide an initial barrier to anything entering. The whole process takes about five minutes in a small wound; it can be slower in wounds with blood flow strong enough to sweep clotting fibrin away.

Coagulated blood has little strength compared to normal tissue, and it is only a temporary expedient to give time in which more permanent measures can be taken. When the skin is breached, surviving cells immediately around the hole of the wound detect that they have no neighbour on that side, and they assemble internal fibres of actin and myosin on the face that no longer has a neighbour. The actin and myosin fibres pull, by a mechanism similar to that covered for muscle in Chapter 6. Because the cells adhere to one another, the whole contractile system produces a 'purse-string' that draws the edge of the hole together (Figure 41). When a wound is very small, for example a puncture with a small hypodermic needle, this purse-string action may be sufficient to close the hole quickly, with no other actions needed except for the natural tendency of cells to re-form mutual adhesions when they meet across the hole.

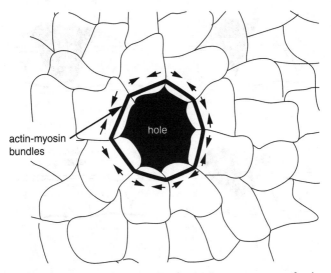

actin-myosin
bundles

hole

41. **Closing of very small wounds by purse-string arrangement of actin and myosin.**

For larger wounds, purse-string closure cannot close the gap, and new material is needed. Damaged cells at the wound edge release factors that attract circulating cells of the immune system such as macrophages. Macrophages engulf pieces of damaged cells and digest them, effectively tidying up the site. They also release further factors that affect neighbouring cells. Collectively, these factors cause the cells at the wound margin to proliferate, cause nearby blood capillaries to sprout new branches, and they attract connective tissue cells—fibroblasts—to enter the wound below the coagulated blood (Figure 42). The fibroblasts proliferate and secrete copious quantities of extracellular proteins, particularly proteins such as collagen that are anyway common in connective tissues such as the dermis. This protein-rich plug forms a platform across which epidermal cells slowly migrate under the overlying scab and meet, sealing over the wound. The meeting is aided by the fibroblasts becoming muscle-like, contractile cells, pulling the

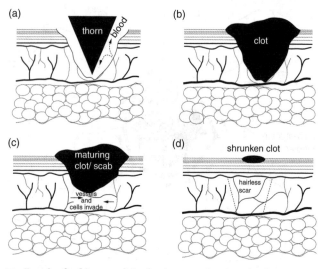

42. **Repair of a skin wound, by clotting, scarring, invasion by epidermis, and shrinkage.**

wound edges together and making the distance the epidermal cells have to travel smaller. Because there are no hair follicles or sweat glands in the filler-tissue the fibroblasts have built, the new epidermis above has neither hairs or pores, and it is usually free of pigment cells as well. This leaves it visible as a scar. The underlying plug is slowly remodelled, with excess blood vessels removed, and hastily assembled extracellular proteins replaced with ones arranged for strength. Nevertheless, scars are not as strong as undamaged skin. Their production is a compromise between the eventual quality of what is made, and the urgent need to restore a barrier.

Fighting micro-organisms

We are surrounded by micro-organisms that would, if they could, literally eat us alive. Most of these organisms are regarded as harmless, but this does not mean that they would politely leave us alone if we were defenceless. Rather, it means that we have evolved defences against them that are completely effective, and that people in good health are at risk only from the tiny fraction of micro-organisms that can slip past our defences, at least for long enough to trigger a temporary illness.

Our first line of defence against anything from the environment is the mechanical barrier provided by the skin, as described above. The next is chemical: the secretions that cover the surfaces of our eyes, the inside of our nose, etc., contain a variety of proteins that attack bacteria. An example is the enzyme lysozyme, found in tears, saliva, milk, and the mucus of the nose and vagina; the enzyme digests the cells walls of bacteria, and kills them. Defensins form another broad class of anti-bacterial proteins present in various body fluids. These chemical defences are made by the human body itself. Another group are made not by our own tissues, but by bacteria that have evolved to have a symbiotic relationship with us. *Lactobacillus* bacteria, for example, live harmlessly in vaginal mucus and secrete chemicals that prevent overgrowth of fungi such as *Candida albicans*.

Within the tissues themselves, microbes encounter further defences that have both chemical and cellular components. Within the blood and fluids that bathe internal tissues are proteins of the complement system which, rather like the proteins of the coagulation system, work in a cascade so that the triggering of relatively few initiator proteins generates a massive activation of the final effectors. The initiator protein of the complement system is always being activated at a low level, but, as long as it is free in the blood, it is inactivated again quickly by other blood proteins. The initiator can bind the surfaces of bacteria and, when it does so, the inactivating proteins can no longer inactivate it. It is therefore able to trigger the cascade of activations that lead, ultimately, to the formation of membrane attack complexes that punch holes in the membranes of bacteria. Other initiator proteins exist, which bind to carbohydrates on the surfaces of bacteria and trigger the same final cascade of complement proteins. The result is a rapid and vigorous attack on these organisms.

In addition to these chemical defences, the body has a set of specialized cell types devoted partly or wholly to defence. Phagocytes, including the macrophages already mentioned in the context of wound healing, acquire their name (phage = 'eater') from their propensity to engulf large objects such as cell fragments and bacteria. Phagocytes can recognize some bacterial components directly. They can also recognize complement so, when the complement initiator protein is bound to a bacterium, it acts as an 'eat me' signal. When phagocytes engulf bacteria, they secrete a cocktail of chemicals. Some of these are toxic to bacteria, and some are signals that increase local blood flow and cause local vessels to become leaky so that fluid containing complement and more phagocytes and other blood cells can enter the tissue. These activities are collectively known as inflammation and cause the characteristic hallmarks of this state: redness and heat (from increased blood flow), swelling (from fluid entering the tissues), and pain (from swelling and also chemical activity). In the middle of all this may be an area of whitish pus, which consists mainly of phagocytes, dead bacteria, and dead human tissue, some killed by bacteria and some killed by the toxic environment produced by the phagocytes. Skin pimples are a classic and highly visible example. Again, there is a compromise between the need not to cause excessive damage, and the urgent need to stop the infection taking hold. Sometimes inflammation is excessive, and causes more damage than the micro-organism itself.

If human cells are highly stressed, either by infection or by a physical or chemical injury, for example a burn from an iron or a drop of acid, they release molecules that are detected by receptors on phagocytes. These receptors trigger the same kind of inflammatory reaction as do bacterial compounds. The point, presumably, is that even if a micro-organism managed to evolve to be unrecognizable by complement or other detection systems, the damage it did to cells would still be enough to trigger a strong defensive reaction.

Adaptive immunity

The defensive systems described above are ancient, in evolutionary terms, and exist in the same or highly related forms in almost all complex animals. Vertebrates, which tend to be large and to live for a long time, have added an extra layer on top of these basic systems which has one feature that they do not: it learns from experience. Because this learning adapts it to the threats the body will meet, it is called the adaptive immune system.

The adaptive immune system depends on the existence of a vast library of T cells, 'T' standing for 'thymus', the gland in the neck in which these cells develop. Each T cell carries a T-cell receptor (TCR), and the TCRs of different T cells are distinct from each other. They all arise from the same basic gene but, as T cells develop, they activate DNA-shuffling enzymes that introduce random mutations into that gene, and thus different T cells acquire different mutations of the gene and make different versions of the TCR protein from that gene. Young T cells that have just finished this DNA rearrangement are surrounded, in their thymus nursery, by cells that present fragments of body proteins on their surfaces. If a T cell's TCR fails to react even a little to any of these, the T cell dies, presumably because its TCR is evidently useless (Figure 43). If a T cell's TCR binds any of the presented proteins very strongly, it also dies, presumably because it is reacting to a normal body protein, which would be inappropriate. But if a T cell's TCR binds presented proteins only weakly, the cell lives, having proved its TCR is capable of working and that it does not react strongly to the body. It then leaves the thymus for a life out in the general blood and lymph circulation. The result of all of this is that the body will have populated itself

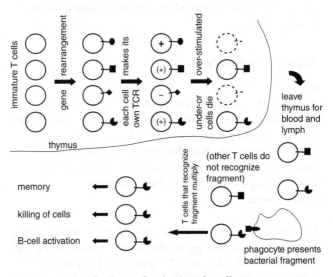

43. Generation, selection, and activation of T cells.

with millions of T cells, carrying millions of different versions of a working TCR that is not strongly activated by any of the body's own proteins.

The T cells frequently encounter phagocytes and, if these phagocytes have recently engulfed anything such as a dead cell or bacterium, they will present digested fragments of it on their surfaces. Most of the time, the TCR of the T cell will not recognize what is being presented and the cells will go their separate ways unaffected by their encounter. Occasionally, though, the T cell's TCR will turn out to be a good fit for the fragment being presented. The TCR activates the T cell, and the cell begins its defensive work. It proliferates quickly, creating a small army of daughter cells that carry the same TCR. Some of these cells develop into specialized 'killer' T cells, and if they encounter any general tissue cell carrying the protein recognized by the TCR, they will kill it. This may seem extreme, but it is a useful way of dealing with micro-organisms that try to evade surveillance by living inside the body's cells. Other T cells with that TCR will remain in the circulation to form a kind of 'memory'—a large number of cells that recognize the sign of this particular infection and that can swing into action very quickly, without delays for proliferation, if it is encountered again. In this way, the body learns about threats met once, and is more prepared to meet them a second time. This is the basis of vaccination: to treat the body with a harmless version of a dangerous micro-organism, or with relevant fragments of it, so that the memory is formed and the body will mount a strong and rapid response if the real micro-organism is ever met.

A third set of the activated T cells interacts with another type of immune cell, B cells (which develop in bone marrow). B cells have their own version of TCRs, called BCRs and, like T cells, B cells generate millions of variants by the same process of DNA shuffling. B cells patrol the body and, if they find a molecule that their BCR recognizes, they take up the molecule, process it, and present fragments of it on their surfaces on a specialized fragment-displaying protein. If a T cell encounters a B cell displaying a fragment that is recognized by its TCR, it makes a signal that activates the B cell. The B cell multiplies to make a large number of daughter cells all with the same BCR. Some of these daughters remain as they are, again to form a memory to speed things up if the same threat is encountered in the future. The others mature into a different type of cell, a plasma cell, that secretes BCR. Secreted BCR is called antibody, and it can spread rapidly through body fluids. When antibodies bind to the surface of microbes, they can activate the complement system by a different pathway, and thus kill the microbe. They too act as powerful 'eat-me' signals, strengthening the response of phagocytes and directing them even to micro-organisms that the antibody can detect but the phagocyte itself cannot.

The adaptive immune system therefore acts as a kind of amplifier of the ancient, innate immune system, and it also brings the new features of learning and memory. But the basic features of having an impenetrable barrier and chemical defences are, while perhaps less glamorous than an adaptive, learning system, still absolutely essential.

Chapter 8
Reproduction

Individual lives are finite. Homeostasis, though impressive, is imperfect. Over time, more and more of our cells acquire genetic, chemical, and structural defects that they cannot correct. Even if we manage to avoid all accidents and infectious diseases, this accumulating population of damaged cells among the healthy ones causes our bodies to age, to develop non-infectious diseases (such as coronary disease, neurodegeneration, and cancer) and, eventually, to die. The long-term continuation of humanity therefore depends, as it always has, on our ability to reproduce in a way that does not pass ageing forwards to the new generation, and that produces new humans that are, on average, at least as good as previous generations at avoiding infectious disease.

The requirement to reset the 'age clock' is met by starting a new body from one single cell. Senescent cells are not capable of doing all they have to do to produce a new baby, so the fact that a baby is being made at all is a kind of quality control step to ensure that the single cell from which it began was in good shape. Added to this, the vast amount of rapid growth that takes place from the starting cell to baby (around 100,000,000-fold in volume) greatly dilutes any chemical contaminants that might be in the cell. Starting anew from a single cell may be an effective method of cheating time and decay, but it is enormously expensive because the whole business of developing the complexity of a human body must be achieved all over again in every generation.

The second requirement, avoiding infectious disease at least as well as previous generations did, is subtle but worth understanding because it has such an influence on the way we reproduce. Micro-organisms reproduce much more quickly than humans, and have generation times as short as twenty minutes rather than our roughly twenty years. They can therefore evolve relatively quickly. If all humans were exactly the same, in the sense that identical twins are the same, then there would be a high risk that a micro-organism might evolve to be perfectly adapted to evade the defences of that human (Chapter 7) and, therefore, of every other human. This effect has been seen in agriculture; the Irish potato famine was so destructive because most of the potatoes grown in Ireland at the time were clones of one another, so when a fungus evolved to prey on that exact potato, it could spread through the whole population of potatoes unchecked. The main defence that successful species have against this is variation. The fact that we are not all the same means that a micro-organism well suited to prey on one person will not be as well suited to preying on his neighbour and one epidemic will not wipe out the whole species.

Variation between people arises from our having different versions (alleles) of our genes, and maintenance of this variation depends on shuffling of alleles during the production of each new individual. Shuffling of genes is the ultimate reason for sex, even if it is not the proximate motivation. Every one of us has two copies of each of our chromosomes (ignoring, for a moment, the complications of chromosomes X and Y). We obtained these copies from two different individuals, our mother and our father, in an act of genetic cooperation that produced the single cell from which our bodies developed. The two versions of each chromosome will often carry different alleles of a gene, and effectively give contradictory instructions ('make blue eyes!', 'no,

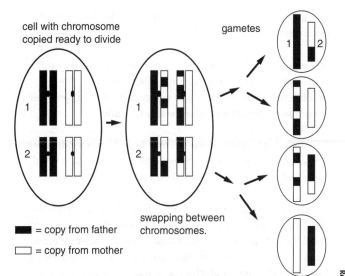

cell with chromosome
copied ready to divide

gametes

1

2

swapping between
chromosomes.

■ = copy from father

☐ = copy from mother

44. **Chromosomal events in meiotic cell division, showing how genetic information from the chromosomes inherited from this individual's father and his mother are mixed in a swapping operation, and passed to gametes that may produce this individual's own child. The diagram depicts just two pairs of chromosomes, though we have twenty-three pairs in all; it also focuses on chromosomal events and ignores other aspects of the cell division.**

brown eyes!'). The body each of us built is effectively a compromise between the rival 'how to build a human' genetic instructions coming from our parents. Given that chromosome number does not increase over the generations, we need to pass on only one copy of each of our chromosomes to our child. In principle, we could do this by passing on either our maternal or paternal version of each but, in reality, we do something more complicated: during the specialized cell division (meiosis) that produces cells with only one copy of each chromosome, parts of the maternal and paternal versions of each chromosome are swapped together to produce a composite (Figure 44). Thus the chromosomes we pass on are a mix of the alleles from each of our parents.

A complete set of chromosomes, one of each, goes into a gamete cell (a spermatozoon or oocyte) and, when these fuse, they produce the single cell (fertilized oocyte) with the full complement of chromosome pairs that will produce the child.

Of course, the sperm and oocyte come from different individuals who are already genetically distinct, and this contributes greatly to mixing of alleles and therefore to variation between individuals.

In humans (and all other mammals), the embryo develops inside its mother's body, where it is fed and protected by her. Even after birth, the baby is fed by mother's milk and looked after for many years by its parents, or sometimes other adults, while it grows up. The physiology of reproduction is therefore very broad, spanning the making of gametes, making a potential home for an embryo in the mother, arranging that fertilization takes place, supporting the developing embryos, making a supply of milk in time for the birth, and, of course, developing from a single cell into an adult capable of repeating the whole process.

Making gametes

The logic of sexual reproduction—making gametes with only one copy of each chromosome and having them fuse together to make a new cell that will go on to be a new individual—does not itself require the gametes to be physically different. In almost all animals and plants, though, an evolutionary instability has meant that they are. The simple logic of evolution means that the organisms whose offspring come to dominate future generations are those that can bequeath as many healthy offspring as possible to each next generation. This creates a pressure for maximizing the number of gametes produced, and in the deep past seems to have driven some individuals toward a strategy of using their available resources to produce more, smaller gametes, carrying fewer general cellular resources with their chromosomes. The increasing prevalence of small gametes created a counter-pressure in other individuals to produce gametes large enough to thrive even if they fused only with a smaller one. The presence of these larger gametes allowed producers of undersized gametes to produce even larger numbers of even smaller ones that would provide far less than their fair share of general cellular resources, and so on.

Zoologists have a special term for this kind of reproductive cheat: 'male'. Males produce very large numbers of very small gametes. Females, on the other hand, produce tiny numbers of very large gametes that provide almost everything an early embryo needs. The great difference can be appreciated from the numbers of gametes produced per year; a woman produces about 13 fully mature oocytes while a man produces about 150,000,000,000 sperm. Much reproductive, and indeed behavioural, biology arises from this difference, often expressed by the phrase 'eggs are expensive, sperm are cheap'.

The production of male gametes is relatively simple, and is a continuous process from puberty to the end of life. The internal structure of a testis is dominated by small, thick-walled tubes called seminiferous tubules. The very outer layer of the tubes consists of muscle-like cells that maintain a gentle peristalsis to move fluid, and any gametes released into it, along. Inside this layer is a layer of stem cells, which divide to produce more stem cells and also cells destined to go on to become sperm (Figure 45). These new cells also divide, many times, to increase their number, and then enter a meiosis to generate four small daughter cells, each carrying just one copy of each chromosome. These cells mature to become sperm. They leave the testis, mainly by passive fluid flow, and are stored in a set of tubes attached to the testis

muscle-like cells

stem cells (just inside muscle-like layer)

rapidly proliferating progeny of the stem cells

cells undergoing meiosis (nearer middle)

maturing sperm (nearest the hollow middle)

45. A cross-section of one of the seminiferous tubules of the testis.

called the epididymis. They are moved to the urethra in a complicated set of smooth muscle contractions during ejaculation, the muscle contractions also causing the release of secretions from glands such as the prostate and seminal vesicles, which provide food for the spermatozoa and also some protection from antimicrobial defence systems in the vagina.

In females, production of gametes is periodic and synchronized with the cycling of the uterus between states in which it is capable of supporting a growing embryo, and states in which it is not. It is not clear whether women show this cycling because there is some deep reason why a uterus cannot be maintained in a receptive state indefinitely, or whether they show it simply because we evolved from ancestors with seasonal breeding patterns. In humans, the cycle has no obvious connection to seasons. It does have a length similar, on average, to that of the lunar cycle (twenty-eight days) but this is probably a coincidence, as many women's cycles are a little shorter or longer than this, and even women with reliable twenty-eight-day cycles do not all synchronize to have the same stage of the cycle at the same phase of the moon.

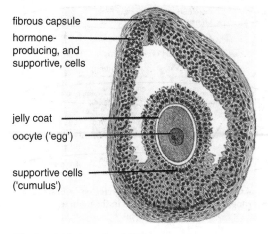

fibrous capsule

hormone-producing, and supportive, cells

jelly coat

oocyte ('egg')

supportive cells ('cumulus')

46. A maturing ovarian follicle.

Without the need to produce vast numbers of gametes, women have no need to maintain a stem cell population in their ovaries and use a very different strategy. Modest stem cell proliferation, subsequent meiosis, and the very beginning of oocyte maturation are completed in foetal life. By the time a girl is born, her ovaries contain a stock of immature oocytes, each surrounded by a layer of supportive cells. The whole oocyte-plus-support unit is called an ovarian follicle (Figure 46). Further development of these follicles is slow at least until puberty. From then on until menopause, groups of follicles begin rapid maturation in response to pulses of follicle stimulating hormone (FSH) from the pituitary gland. These pulses occur once every menstrual cycle, but the process of maturation takes a few weeks, so the menstrual cycle that triggers follicles to resume maturation is not the one in which they will be released for possible fertilization. Most of the group will not, in fact, be released; some will develop at the wrong pace to be at precisely the correct stage for the relevant menstrual cycle and will die, while the small group that are correctly timed undergo a

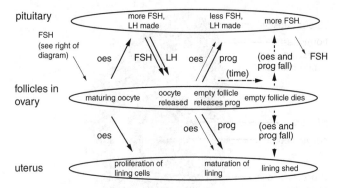

47. A summary of the events of the menstrual cycle: 'oes' = oestrogen, 'prog' = progesterone, FSH and LH as defined in the main text.

competitive process from which only one winner emerges to complete its maturation.

The support cells in the follicles make hormones that report the state of the follicle's maturation to the uterus and to the pituitary. A maturing follicle secretes oestrogen, which causes the cells lining the uterus to proliferate (Figure 47). The rising levels of oestrogen also cause the pituitary gland, on the underside of the brain, to release another pulse of FSH (which will induce fresh groups of immature oocytes to begin to prepare for a future cycle, as well as help promote maturation of the ones for this cycle) and to release luteinizing hormone (LH). This combination of FSH and LH drives the maturing follicle to move to the edge of the ovary and burst, releasing its oocyte into the oviduct (also called the fallopian tube). Some women feel this release as the *Mittelschmerz*—the 'middle pain' of a menstrual cycle—while others are unaware of it. The empty remains of the follicle now produce progesterone, a hormone that drives the now thickened uterine lining to mature, ready for an embryo. As this is happening, the oocyte is carried by fluid flow along the oviduct towards the uterus.

If copulation has taken place around this time, the oocyte may meet and perhaps be fertilized by sperm in the oviduct, and will progress towards the uterus as a young embryo. If fertilization does not take place, the oocyte's moment will be lost, and the cells of the follicle that released it will regress and stop producing oestrogen and progesterone. As the levels of these hormones collapse, the uterine lining will break down and, with the assistance of often painful contractions of uterine muscles, will be lost as menstrual flow. The pituitary responds to the hormonal fall by secreting FSH, which will lead the follicle that is ready for the next cycle to enter its final phase of maturation, and the cycle repeats.

Fertilization and implantation

Fertilization is a complicated process, involving sperm having to digest their way through a jelly-like protective coat around the oocyte, and then to fuse their membranes with the membrane of the oocyte. This fusion allows their chromosomes to enter the oocyte and to join them in the first cell division of what is now the embryo. One essential feature of fertilization is that the first-arriving sperm triggers a sequence of rapid changes to the oocyte's surface to prevent any further sperm fusing with it, an event that would add surplus chromosomes and prevent normal development.

The first days of embryonic life are dominated by cell divisions, turning the large oocyte into a ball of identical cells. Once there are 32–64 of these cells, the ones on the outside of the embryo become specialized to form a closed sheet, the trophectoderm. The trophectoderm interacts with the surface of the uterus, sticking to it and invading it. It will form the foetal side of the placenta, while cells of the uterine lining form the maternal side. Later, as the embryo grows and becomes a foetus, it sends blood vessels to the placenta, as does the mother, and these run close enough to one another that oxygen and food can cross from mother to growing foetus. Similarly, waste products, such as those that would pass into the urine in an adult, pass from the blood system of the foetus to that of the mother, who disposes of them in her own urine. Hormones also pass between mother and foetus, and help to synchronize the events of pregnancy, including changes in maternal metabolism to meet the needs of the foetus.

Meanwhile, the clump of inner cells that did not take part in producing the trophectoderm undergoes its own sequence of development, first producing a flat two-layered disc of cells, then turning it into a three-layered structure by a complicated cellular choreography called gastrulation. A sequence of growth and folding events, a little like origami but with cutting and joining being allowed, folds the three-layered disc over to make an outer cylindrical structure, corresponding to the basic roughly cylindrical structure of the body, with two tubes running from head to tail inside it. One of these tubes, eventually open at both ends, is the gut while the other, eventually closed everywhere, is the central nervous system. This basic vertebrate body plan—two tubes running inside an outer cylinder—is then elaborated as it grows by a long sequence of events far too numerous to describe here. The result is the emergence of a human form, albeit with the odd proportions of a foetus rather than an adult.

Sex determination

In humans, biological sex is determined genetically. In addition to having twenty-two pairs of normal chromosomes (autosomes), we as a species have one 'pair' of sex chromosomes, X and Y. X chromosomes have a size and composition typical of chromosomes in general, and we all have at least one. Women have a pair, whereas men have one X chromosome and one Y chromosome, a very small structure containing little in the way of genes. All oocytes contain one X chromosome, but each sperm can contain an X or a Y, thus giving a 50 per cent chance that an embryo will be XX and a 50 per cent chance it will be XY (by the time of birth, the ratios are not precisely 50/50 because there is a genetic risk of having only one copy of genes on the X chromosome, and therefore of XY embryos failing due to a genetic problem; even after birth, this risk gives boys and men an elevated risk of some genetic conditions, such as haemophilia). When an embryo is developing its gonads, certain cells in the gonad 'try' to activate expression of the SRY gene on the Y chromosome. If the Y chromosome is present, SRY will be activated and those cells will be driven down the path of blocking ovary development and building a testis. The developing testis communicates its presence to the rest of the body by secreting hormones (testosterone and anti-Mullerian hormone), and these hormones cause the as-yet ambiguous genital, skeletal, and other tissues that are different in males and females to develop along male lines. If SRY is not present because there is no Y chromosome, ovary development will take place instead and the tissues of the body will develop along female lines.

The fact that only the gonads evaluate the presence of a Y chromosome, and communicate the result to the rest of the body by hormones, has interesting consequences. If for genetic reasons these hormones cannot be produced or detected properly, or if the mother ingests substances that mimic or interfere with these hormones, the body may not develop along the same sexual trajectory as the gonads. An extreme case of this is complete androgen insensitivity syndrome, in which testosterone cannot be detected and people develop bodies that are externally female but have internal testes where ovaries would be. There are also many intermediate states, usually classified under the term 'intersex', although for many instances that term is misleadingly simple. In any case, it is clear that speaking of just two sexes does not cover even the anatomical aspects of all human life.

Some people feel, emotionally, that they are a sex other than the one their body makes them seem to be. It is not clear whether this reflects a decoupling of brain development from the hormonal influences of developing gonads, or whether other influences (from molecular to social) are more relevant, and the answer may be different for different people. It is also clear that sexual orientation—which people, if any, someone finds attractive—is not determined by biological sex. Research by zoologists into the sexual behaviour of other animals suggests that a spectrum of individuals and behaviours is not an unusual feature of mammalian life; the possible evolutionary adaptive value of having such a spectrum is a much-debated topic.

Birth

Compared to our closest relatives in the animal kingdom, human babies are born very immature, probably because we as a species have large heads for our body size. The route that the vagina takes between uterus and vulva takes it through the lower opening of the pelvis, which limits the size of a head that can pass through it to around 10 centimetres diameter (about the size of a DVD). This means that a baby must be born before its head grows larger than this limit, even if this means being born particularly helpless.

The timing of birth is set by a combination of interacting hormonal systems in mother, foetus, and placenta. During most of pregnancy, progesterone acts as a powerful block on the events that will eventually drive birth: contraction of the uterine wall muscles, and relaxation of the uterine cervix. From around the end of the 8th month of pregnancy, an increase in progesterone sequestering proteins in the placenta and a decline in the sensitivity of the uterine muscles to the hormone allows the uterus to begin to contract, gently at first but more forcefully as the month goes on. At about the same time, tissues derived from the foetus and those on the maternal side of the placenta secrete prostaglandins, which are powerful drivers of uterine muscle contraction and also of the cervix's preparation for opening.

Oxytocin, a hormone released by the pituitary glands of both the foetus and the mother, is a critical hormone in labour and giving a mother extra oxytocin is one method of inducing labour to begin. At least in experimental animals, where experiments can be done that cannot ethically be done with humans, oxytocin is important in recognition and bonding between mother and child (physiologists tend to avoid words like 'love', but that is essentially what is meant: some aspects of sexual love, too, involve oxytocin).

For the mother, giving birth involves a great deal of uterine contraction, massive dilation of cervix and stretching of the vagina, and loss of at least some blood as the placenta falls away. Obviously, these deformations of tissue are traumatic and usually involve a great deal of pain, which may be controlled pharmacologically according to availability and the mother's wishes. Often, such control is mediated locally to a specific area of the mother's spine, to avoid exposing the foetus to drugs. It is a reasonable assumption that the process of birth is traumatic and painful to the foetus too, but it occurs before our brains lay down memories we can access in the normal way as adults. As well as going through the mechanical process of being expelled from the uterus, the foetus has to cope with very rapid changes to its physiology as the placental functions fail. During a normal birth, the foetus is born before the placenta is expelled, so at least some oxygen exchange between mother and foetus continues. But as soon as the baby, as it is now called, is in the air it needs to begin breathing for itself. This does not just involve expelling amniotic fluid from the lungs and inflating them for the first time, but also making a major alternation of blood flow.

As explained in Chapter 2, in a free-living human, 'used' blood flows through the right side of the heart to the lungs to collect oxygen, then to the left side of the heart, then back to the rest of the body. In a foetus, the lung part of the circulation is pointless, so most of it is 'short-circuited' via two direct connections. One, the foramen ovale, is a hole that cross-connects the upper chambers of the heart, and the other, the ductus arteriosus, is a pipe that connects the pulmonary artery to the aorta. These are excellent adaptations to foetal life but would interfere with lung function if they persisted after birth. The ductus arteriosus remains open only as long as it receives high levels of specific types of prostaglandins that are made in the placenta. When the connection to the placenta is lost, smooth muscles in the wall of the vessel contract and close it off. Once the lungs inflate after birth, the changing back-pressure from them alters the pressure in the right side of the heart and this deforms the tissue enough to close the foramen ovale immediately; it then seals over within a few weeks.

Babies need to make no adaptation for excretion—their kidneys are already active in foetal life and produce the shock-absorbing amniotic fluid that surrounds them in the uterus (the 'waters' that 'break' early in labour). But they do need to adapt to a new source of food, one unique to mammals: milk.

Lactation and infant care

Puberty in girls is marked not just by onset of follicular maturation and menstrual cycling, but also by the development of mammary glands: essentially sweat glands that have evolved to secrete a liquid that gives both food and antibodies to a baby. The pre-pregnant mammary gland contains a tree-like arrangement of ducts that ramify through fatty tissue and converge on the nipple. The tree expands in the long, progesterone-rich period of early pregnancy and, under the influence of this and other hormones such as prolactin (PRL), the ducts develop rounded side-branches called alveoli. The cells lining these develop the capability of secreting milk, which they begin to do towards the end of pregnancy. By birth, there is usually enough of this early-type milk (colostrum), rich in defensive cells and proteins, to feed the newborn baby ('usually' because some women do not produce enough milk, and feeding has to be done another way). After birth, the loss of placental hormones changes the composition of the milk less toward defence and more toward nutrition. Suckling

128

by the child maintains levels of PRL, so that milk production is usually maintained for as long as there is demand. Suckling also suppresses pituitary release of FSH and LH, thus delaying the resumption of menstrual cycling and the possibility of another pregnancy (though the length of this delay varies between women and between pregnancies, so it is not a reliable method of contraception). When breast-feeding ceases, the alveoli regress. New ones will form in the event of another pregnancy.

Many processes of development begun in the womb continue after birth; growth, obviously, but also completion of the bony skeleton and a great deal of refinement of the nervous system. Importantly, neural development is controlled not only by genetics but by the activity of the axons and synapses, much of which is driven by the environment. Our vision, for example, is initially crude and we 'learn' to see properly as the connections within eyes, and connections to and within the visual centres of the brain, are refined once a baby starts to view the world. Learning to crawl and then walk is broadly similar: once the musculoskeletal system is strong enough to allow movement, babies begin to move clumsily but learn by their errors to move in more and more efficient ways. This facility for learning new motor patterns continues throughout life, and even as adults we approach completely new tasks clumsily with a need to think of every individual action, until with practice the sequences of movements become automatic. Anyone who has learned to drive or to dance or to play a musical instrument will be familiar with the sequence of events.

Much of what it is to be human derives not directly from physiology, but from culture. Our physiology gives us the ability to make a complex range of sounds, but it is our culture that gives them meaning as we tie them together into language. The physiology of our brains allows us to learn, but our culture allows what has been learned by others to be passed on without having to be discovered anew. A great deal of what is important in human reproduction therefore lies not in raw biology, but in a childhood filled with education in how to live as part of the tribe, nation, or broader culture. These things are outside physiology as a discipline, but are mentioned here because it would be impossible to produce complete new human beings without them.

Choices

One of the key themes of physiological control, which has featured in many chapters of this book, is homeostasis: an avoidance of dangerous excess. But this does not apply to reproduction, at least not in the sense of applying to its results. In non-human animals, young are over-produced and compete for resources so that only the fittest and most fortunate survive to produce young of their own. This is the driving force of natural selection; in Darwin's own words, 'One general law, leading to the advancement of all organic beings, namely, multiply, vary, let the strongest live and the weakest die' (*The Origin of Species*, 1859). This remains true for many humans; a large proportion of children on this planet are born to a world of famine, drought, disease, and war (which usually has something to do with resource allocation, even if combatants pretend otherwise). Vast numbers die without reaching adulthood.

Our scientific understanding of reproductive physiology now gives us choices. We can choose to limit our reproduction by measures ranging from non-technical (e.g. sexual abstinence) to simple barriers to fertilization (e.g. condoms) to sophisticated manipulation of hormonal signalling (e.g. oral contraceptive pills) to political (e.g. state support in old age to remove a common reason to produce a large supportive family). Yet few societies choose to maintain populations at levels that allow their long-term future to be sustained. Those in the developed world have been lucky that the speed of technical advances has provided short-term solutions in agriculture, medicine, and energy that have kept pace with our growing population, but willingness to use these short-term solutions has only made the underlying problem worse. Soils are being exhausted, atmospheric carbon dioxide is rising, fresh water is becoming less easy to obtain, antibiotic resistance is threatening to reverse half a century of progress against infectious disease, and our economic systems still depend on continual 'growth'.

Our highly evolved brains, and scientific society, now give us an ability to extend the realm of physiological principles from internal matters to external, using them to mediate homeostatic control at an ecological level. Put simply, we can choose to limit our numbers to what can be sustained, long term, by the renewable resources around us. Whether we choose to do so, or fall back into the animal world of a struggle for survival, a struggle that most lose long before they have the chance to become old, remains to be seen.

Concluding remarks

Human physiology is a vast subject, both in terms of the number of systems to be studied, and in the number of fine details that exist in the mechanisms of almost all of them. It is inevitable, therefore, that a short introductory book such as this can present only a tiny subset, chosen to explain the actions of familiar body systems and also to illustrate a range of important physiological principles.

One principle that should have shone through is the emergence of control through the cooperative, collective action of different entities. This point is being emphasized because there has been a tendency in recent years, driven perhaps by metaphors such as 'the selfish gene', to view biology as hyper-competitive at all levels. All of the proteins mentioned in this book—the enzymes, the actin, the myosin, the receptors, the protein signals, the ion channels, and more—are made by genes. Yet the gene for any one of them, in the absence of the others, cannot reach another generation. It is only by working together, in a connected system, that the genes, the proteins made from the genes, the cells made from the proteins, and the tissues made from the cells, can be part of a healthy living system that can reproduce and become part of evolution's story. There are contexts in which it is correct to describe genes, or more correctly different alleles of a gene, as being in competition for dominance in a species. But the means of winning that competition is largely through the proteins encoded by the allele of the gene cooperating so well with the other proteins of the body that some advantage is given to the individual as a whole.

Studying physiology is about more than gaining facts; it is also about gaining an outlook that can see, analyse, and appreciate the action of cooperative systems. It is an outlook that fully acknowledges the importance of understanding at the molecular level, but that can still manage to 'see the wood for the trees' and retain a focus on what the molecules are *for*. Physiology is, indeed, one of the few areas of natural science in which researchers use the word 'purpose' without drawing quote marks in the air as they say it. The same is true, when sensory and cognitive neural systems are being discussed, for the word 'meaning'. In this respect, as well as in its core subject material, the science connects more than most with our deepest questions about what it is to be human.

Further reading

Richard, W. (2019) *The human body book: an illustrated guide to its structure, function and disorders.* Dorling Kindersley. Do not be put off by the 'child-friendly' publisher—this book is so good I recommend it to new medical students as a way to get rapid and accurate orientation in a new topic before diving in to more detailed texts.

Bryson, B. (2019) *The body: a guide for occupants.* Doubleday. Humorous and accurate.

Miller, J. (1980, reprinted 2000) *The body in question.* Pimlico. Jonathan Miller's erudite and interesting account of human physiology from the point of view of a doctor rather than a scientist: much excellent historical background.

Monamy, V. (2017, 3rd edition) *Animal experimentation, a guide to the issues.* Cambridge University Press. A longer discussion about the ethical and scientific issues around animal experimentation, discussed briefly in Chapter 1 of this Very Short Introduction.

Klenerman, L. (2015) *Human anatomy: a very short introduction.* Oxford University Press. Provides information on the anatomy that is linked closely to human physiology, especially with respect to movement.

Davies, J. A. (2015) *Life unfolding: how the human body creates itself.* Oxford University Press. This book considers how humans develop, from egg to adult.

Noble, D. (2008) *The music of life: biology beyond genes.* Oxford University Press. This book explores further the theme alluded to in the concluding remarks, about the different perspectives offered by a systems-centred view instead of a gene-centred one.

Index

For the benefit of digital users, indexed terms that span two pages (e.g., 52–53) may, on occasion, appear on only one of those pages.

Index

T

T cells 111–14
taste 69
temperature 36
tendons 96
thermoregulation 36–7
3Rs 12
tone (of muscles) 98
tongue 69
touch 58–62
triceps 96–7

U

ultrasound 8–9

V

veins 5–7
villi 25–6
vision 69–75
vitamins 26–7

W

Wiener, Norbert 42–4
witness 74–5
wound healing 105–10

X

X-rays 8

CONSCIENCE
A Very Short Introduction
Paul Strohm

In the West conscience has been relied upon for two thousand years as a judgement that distinguishes right from wrong. It has effortlessly moved through every period division and timeline between the ancient, medieval, and modern. The Romans identified it, the early Christians appropriated it, and Reformation Protestants and loyal Catholics relied upon its advice and admonition. Today it is embraced with equal conviction by non-religious and religious alike. Considering its deep historical roots and exploring what it has meant to successive generations, Paul Strohm highlights why this particularly European concept deserves its reputation as 'one of the prouder Western contributions to human rights and human dignity throughout the world.

www.oup.com/vsi

EPIDEMIOLOGY
A Very Short Introduction
Rodolfo Saracci

Epidemiology has had an impact on many areas of medicine; and lung cancer, to the origin and spread of new epidemics. and lung cancer, to the origin and spread of new epidemics. However, it is often poorly understood, largely due to misrepresentations in the media. In this *Very Short Introduction* Rodolfo Saracci dispels some of the myths surrounding the study of epidemiology. He provides a general explanation of the principles behind clinical trials, and explains the nature of basic statistics concerning disease. He also looks at the ethical and political issues related to obtaining and using information concerning patients, and trials involving placebos.

www.oup.com/vsi

GENIUS
A Very Short Introduction
Andrew Robinson

Genius is highly individual and unique, of course, yet it shares
a compelling, inevitable quality for professionals and the general
public alike. Darwin's ideas are still required reading for every
working biologist; they continue to generate fresh thinking
and experiments around the world. So do Einstein's theories
among physicists. Shakespeare's plays and Mozart's melodies
and harmonies continue to move people in languages and
cultures far removed from their native England and Austria.
Contemporary 'geniuses' may come and go, but the idea of
genius will not let go of us. Genius is the name we give to a quality
of work that transcends fashion, celebrity, fame, and reputation:
the opposite of a period piece. Somehow, genius abolishes
both the time and the place of its origin.

www.oup.com/vsi

MEMORY
A Very Short Introduction
Michael J. Benton

Why do we remember events from our childhood as if they happened yesterday, but not what we did last week? Why does our memory seem to work well sometimes and not others? What happens when it goes wrong? Can memory be improved or manipulated, by psychological techniques or even 'brain implants'? How does memory grow and change as we age? And what of so-called 'recovered' memories? This book brings together the latest research in neuroscience and psychology, and weaves in case-studies, anecdotes, and even literature and philosophy, to address these and many other important questions about the science of memory - how it works, and why we can't live without it.

SLEEP
A Very Short Introduction
Russell G. Foster & Steven W. Lockley

Why do we need sleep? What happens when we don't get enough? From the biology and psychology of sleep and the history of sleep in science, art, and literature; to the impact of a 24/7 society and the role of society in causing sleep disruption, this *Very Short Introduction* addresses the biological and psychological aspects of sleep, providing a basic understanding of what sleep is and how it is measured, looking at sleep through the human lifespan and the causes and consequences of major sleep disorders. Russell G. Foster and Steven W. Lockley go on to consider the impact of modern society, examining the relationship between sleep and work hours, and the impact of our modern lifestyle.

www.oup.com/vsi

THE HISTORY OF MEDICINE
A Very Short Introduction
William Bynum

Against the backdrop of unprecedented concern for the future of health care, this Very Short Introduction surveys the history of medicine from classical times to the present. Focussing on the key turning points in the history of Western medicine, such as the advent of hospitals and the rise of experimental medicine, Bill Bynum offers insights into medicine's past, while at the same time engaging with contemporary issues, discoveries, and controversies.

www.oup.com/vsi